Pertinent and Traditional Approaches Towards Fishery

Edited by Noor Us Saher

Published in London, United Kingdom

Pertinent and Traditional Approaches Towards Fishery
http://dx.doi.org/10.5772/intechopen.94733
Edited by Noor Us Saher

Contributors
Giang Van Tran, Quang Minh Dinh, Yaisel Juan Borrell Pichs, Marina Parrondo Lombardía, Lucía García-Florez, Eduardo Dopico Rodríguez, Zardasht Taha, Mullaivanam Ramasamy Sivakumar, Rangasamy Shanthi, Toru Yazawa, Noor Us Saher, Noor Hawa, Abdul Hameed Baloch, Muhammed Tabish, Mussarat ul Ain, Faiqa Razi, Naureen Aziz Qureshi, Raoof M. Niazi, Altaf Hussain Narejo

First published in London, United Kingdom, 2023 by IntechOpen
IntechOpen is the global imprint of INTECHOPEN LIMITED, registered in England and Wales, registration number: 11086078, 5 Princes Gate Court, London, SW7 2QJ, United Kingdom

British Library Cataloguing-in-Publication Data
A catalogue record for this book is available from the British Library

Additional hard and PDF copies can be obtained from orders@intechopen.com

Pertinent and Traditional Approaches Towards Fishery
Edited by Noor Us Saher
p. cm.
Print ISBN 978-1-80355-204-0
Online ISBN 978-1-80355-205-7
eBook (PDF) ISBN 978-1-80355-206-4

Meet the editor

Dr. Noor Us Saher is an active researcher in marine biology and a head of the Centre of Excellence in Marine Biology, University of Karachi, Pakistan. She obtained a Ph.D. in Marine Zoology from the same university in 2008. Dr. Saher has eighty journal articles and six book chapters to her credit. She is also the author/coauthor of about 400 various nuclear and mitochondrial papers including novel genes of marine species. She has executed seven academic and research projects as a principal investigator and three projects as a research officer. She has supervised nine Ph.D. and four MPhil scholars to date. Dr. Saher received the 2023 Dr. Nasima M. Tirmizi Gold Medal from the Pakistan Zoological Society.

Contents

Preface

Fish and fisheries play a significant role in ecosystems around the world. Seafood is a source of healthy dietary protein and the seafood industry brings in more than a trillion dollars annually and employs millions of people. Fish stocks are a key component of the global food system and a vital part of the aquatic environment and food chain. According to the Organization for Economic Co-operation and Development (OECD), the ocean's economic output is $1.5 trillion annually; this number is expected to increase to $3 trillion by 2030. The international trade value of fish is approximately $40 billion per year. Fisheries raise and harvest fish and other aquatic life and are sites for fishing activities. Commercial fisheries include wild fisheries and fish farms, both in freshwater areas (about 10% of all catch) and the oceans (about 90% of all catch). About 500 million people worldwide are economically dependent on fisheries. In 2016, 171 million tons of fish were produced; however, overfishing is an increasing problem causing declines in some populations.

Healthy oceans are not only a source of jobs and food but are also important for economic growth, climate regulation, and the interests of coastal communities. Proper knowledge of the life, habits, and behavior of fish is essential to plan, control, and manage fishery resources. Monitoring and analyzing fish diversity, abundance, and catch, as well as target and non-target fishing, types of capture methods, and fishing fleet, provides information about the biological sustainability of commercial fishing. Catch data measured in stock assessments is helpful to determine sustainable fishing.

Fish and fisheries play critical roles in ocean health, community well-being, food security, and economic stability. Ensuring their sustainability is a key goal of governments, civil society, philanthropies, and development institutions. Collectively, this work covers a wide variety of types of fisheries, affected species, habitats, ecosystems, jurisdictional and policy contexts, management capabilities, and socioeconomic conditions. The application of scientific knowledge for the development of the fishing industry lies in an intimate knowledge of the biology of fish.

Fisheries science plays a crucial role in understanding and managing aquatic ecosystems and the resources they provide. Fishery scientists study aquatic ecosystems, including the physical and chemical environment, and the interactions between species and their habitats. This helps to provide a comprehensive understanding of these systems and their ecological processes.

This book summarizes different directional studies on fishery biology and sciences. It highlights the role of fisheries in the biological and management sciences through the use and application of various traditional and advanced biological approaches. It is organized into two sections. Section 1, *Traditional (Morphological and Ecological) Approaches*, discusses some traditional approaches and focuses on the relevance and importance of morphology, ecology, and biological studies. Section 2, *Advanced (Cellular and Molecular) Approaches* focuses on the importance of advanced cellular and molecular studies in the fishery sciences.

The taxonomy or identification of species is the way in which living organisms are named, described, and classified based on multiple approaches ranging from morphological descriptive characters to higher molecular levels. Using morphological observations, taxonomists identify, describe, and arrange the species into families, classes, phylum, and other divisions. Chapter 1, "Perspective Chapter: Crustaceans Taxonomy", discusses crustacean (crab, shrimp, lobster, and crayfish) taxonomy and classification, as this diverse and important group of fish, has economic implications worldwide.

There is an intimate relationship and interaction between the aquatic ecosystems, including the physical and chemical environments, species, and their habitats. This not only helps to provide a comprehensive understanding of these systems and their ecological processes but also reveals the impact of environmental conditions and quality on the overall biodiversity of the aquatic body as well as species morphology, biology, and other related traits. Chapter 2, "Eco-Morphology of Some Decapod Crustaceans in a Tropical Coastal Marine Waters", highlights this relationship between the morphology, population structure, and dynamics of crustaceans and the water quality of the coastal areas of Ohio Nigeria.

The study of fish populations and their habitats emphasizes the sustainability of biodiversity on the coasts as well as determines sustainable fishing levels. The knowledge of the physical and chemical factors (tides, seasonal variations, dissolved oxygen, nutrients) important for fish presence and survival as well as for their spatial and temporal distributional pattern, natural history, and ecology of organisms affecting the particular fishery play a crucial role in understanding and managing aquatic ecosystems and the resources they provide. Chapter 3, "Biological Characteristics of Some Fish Species in the Mekong Delta, Vietnam", provides information about the variety of freshwater and brackish water fish in the Mekong Delta river system. The chapter discloses the fish population structure, distribution, species composition, and their roles in the environment of Vietnam waters. In addition, it examines the growth patterns of some commercial fish and their fishing status.

The provision and availability of food are important for the presence, existence, and growth of fish species. Chapter 4, "Feeding Diversity of Finfish in Different Wild Habitat", reveals the distribution pattern of finfish through grouping (tidal visitors, permanent residents, partial residents, and seasonal visitors) and its relationship with food and feeding behavior. The chapter describes the mode of feeding in most fish families and discusses the various tools and techniques of analytical research for feeding ecology.

Achieving sustainability in fisheries is a focus of fishery biologists and scientists. Anthropogenic, environmental, and climatic changes have drastic impacts on aquatic bodies. As a result, water bodies face challenges such as changes in ecosystems, species habitats, distribution, and dispersion. Fishery biologists use both traditional and advanced tools to guide management and conservation efforts, ensuring that fish populations remain sustainable, healthy, and productive. Chapter 5, "Sustainable Management Plans in Fisheries and Genetic Tools: An Overview of the Challenge in Invertebrates' Fisheries at the Central Area of the Southern Bay of Biscay, Spain," provides an overview of the challenges faced by invertebrate fisheries in the southern Bay of Biscay, Spain. It also examines the use of genetic markers as important tools for understanding the ecological and genetic status of commercial species as well as the management and planning of fisheries.

The latest developments in fish biology research include fish nutrition, ecology, reproduction, fish development, fish immunology, biomedics, and fish welfare. This information is helpful for species adaptability, sustainability, and restoration and conservation management. Chapter 6, "Plasma and Hemocyanin Phenoloxidase Derived from the Hemolymph of Giant Freshwater Prawn *Macrobrachium rosenbergii* (De Man, 1879)" highlights the importance of hemocyanin in triggering innate immunity through one of its subunits that functions as the active moiety. It describes the biochemical composition and other molecular constituents of particular proteins to provide insight into the biochemical approaches in the fishery.

Chapter 7, "Isolated Crayfish Stretch Receptor Neuron Electrophysiology May Explain a Longstanding Mystery of Human Brain Functioning: Eureka Moment", uses the example of crayfish to help resolve the mystery of human brain functioning. It focuses on the complex system of neuron functions by describing single-neuron physiology (Eureka moment), which can be helpful to understand the privileged function of the human brain.

I am indebted to many people who have contributed to the successful publication of this book. First, I would like to thank the authors for their excellent chapters. Their cooperation and patience are highly acknowledged. Second, I would like to thank the staff at IntechOpen, particularly Author Service Managers Sandra Maljavac, Andrea Tomurad, Sara Tikel, and Paula Gavran and Commissioning Editors Jelena Germuth and Filip Lovricevic.

Dr. Noor Us Saher
Centre of Excellence in Marine Biology,
University of Karachi,
Karachi, Pakistan

Section 1

Traditional (Morphological and Ecological) Approaches

Chapter 1

Perspective Chapter: Crustaceans Taxonomy

Zardasht Ahmed Taha

Abstract

Crustaceans are one of the organisms living on the earth that is important to know forms (e.g., larvae, morphs, adults, sexes) and units. In the current era, crustaceans (crab, shrimp, and lobster) emerge as one of the most demanding seafood than other types of seafood, and therefore, natural implication of worldwide fisheries is more dominated by crustaceans. Because a large number of crustaceans taxa have been identified and described, it is necessary to definite groups. Crustaceans are one of the world's oldest and most diversified arthropods, as well as one of the most successful invertebrate groups, with an assessed 40,000 living species and more than 150,000 identified. The names "shrimp" and "prawn" do not relate to any specific taxonomic groups. Although "shrimp" is often used to refer to smaller species and "prawn" is more commonly used to refer to larger ones, there is no clear distinction between the two terms, and their usages are sometimes misunderstood or even reversed in various regions. This chapter aims to focus on the taxonomy of the crustacean and their contribution and importance in fisheries.

Keywords: crustaceans, taxonomy, classification crustaceans, binomial nomenclature, nomenclature rules

1. Introduction

Crustaceans are a varied group of arthropods (phylum Arthropoda) that includes Brine shrimp, copepods, ostracods, Sand hoppers, crayfish, crabs, lobsters, barnacles, prawns, krill, and mantis shrimp (see **Figure 1**) [1]. Crustaceans have branched (biramous) appendages, chitin and calcium exoskeleton, two pairs of antennae projecting in front of the mouth, and paired appendages that function as jaws with three pairs of biting mouthparts. Carcinologists have long argued the taxonomic classification of crustaceans, with five, six, or even 10 classes sometimes assigned to the group at the phylum, subphylum, or superclass level [2]. The Crustacean has been classified as a class by others [3].

2. Taxonomy, systematic, and classification

The term taxonomy is derived from the Greek *taxis* (it means "order" or "arrangement") and *nomos* (it means "law" or "science") (A.P. de Cadolle, 1813).

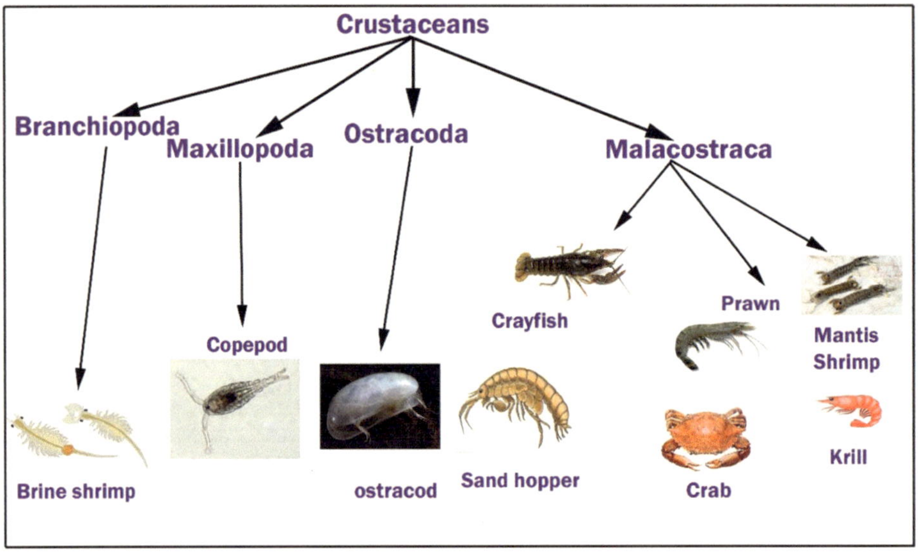

Figure 1.
Illustration of the major groups of freshwater crustaceans [1].

Class Branchiopoda	Class Ostracoda	Class Maxillopoda	Class Remipedia
Subclass Phyllopoda	Subclass Myodocopa	Subclass Tantulocarida	Order Nectiopoda
Order Diplostraca	Order Myodocopida	SubclassMystacocarida	Family Godzilliidae
Suborder Cladocera	Order Halocyprida	Subclass Copepoda	Family Micropacteridae
Family Daphniidae	Subclass Podocopa	Order Platycopioida	Family Speleonectidae
Family Moinidae	Order Platycopida	Order Calanoida	
Family Bosminidae	Order Podocopida	Order Misophrioida	
Family Macrothridicae	Subclass Mystacocarida	Order Cyclopoida	

Table 1.
Crustacean taxonomy is constantly being revised and reevaluated, and the classification system utilized herein is no exception (generally based on [3]).

Taxonomy is the rehearsal and application of classification and can be used ass taxonomic units also familiar as taxa. It originally can be denoted only to the classification of organisms but up-to-date it is used in a fuller sense to the principle fundamental such as a classification. Taxonomy essentially copes with the naming and classification of taxa and forms only a part of biological systematics known as biodiversity. In taxonomy, the smallest unit of classification is called species, which consists of classification and nomenclature. Based on the relation of the organisms, it can be grouped or placed known as classification [4]. The system of classification is developed and used by Carl Linne, 1735, the term systematic can be used in taxonomy.

Crustaceans are divided into six categories (see **Table 1**).

1. Ostracoda—tiny animals with bivalve shells [5, 6].

2. Branchiopoda—*clam shrimp.*

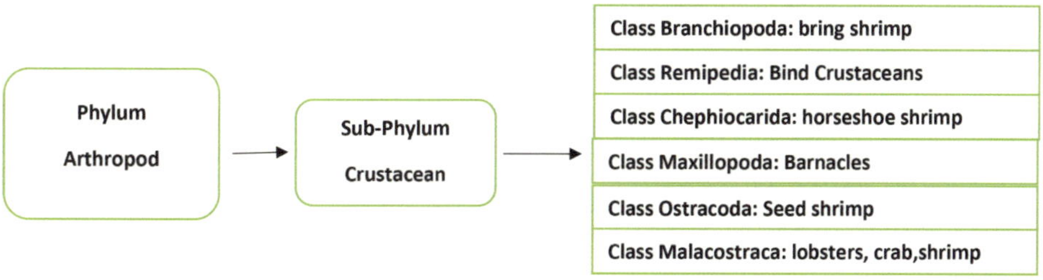

Figure 2.
Crustaceans groups.

3. Cephalocarida-horseshoe shrimp.

4. Maxillopoda—include barnacles.

5. Malacostraca—*great diversity.*

6. Remipedia—blind crustaceans.

Crustaceans are sometimes separated into two groups: entomostracans, which include copepods, barnacles, and fairy shrimp, and malacostaceans, which include lobsters, crabs, and shrimp (**Figure 2**).

3. Species of crustaceans

Crustaceans consisting of crabs, shrimps, and lobsters are one of the most important groups of Arthropod [7]. Now can be explained order Decapoda, for the first time about 8000 species of crustaceans dispersed among more than a thousand genera (Chace, 1951) and lately estimated a total number of extant species of 14,756 in 2725 genera (see **Figure 3**). This implies that in the last 50 years, the number of described species has nearly doubled. However, we are a long way from knowing the true global diversity of decapods. The number of known fossil species currently stands at 3300, and discoveries will continue as new localities are explored, more revisions are completed, and museum collections are more thoroughly studied.

According to Ernst Mayr's description of a species was groups of essentially or hypothetically upbringing natural populations, which are reproductively isolated from other groups (see **Table 2**). Moreover, for accurate taxonomic species, there are numerous kinds relating to ecological and evolutionary concepts (e.g., sibling species, cryptic species, sympatric species, allopatric species, syntopic species, allotopic species, monotypic species, insular species, polytypic species, continental species).

3.1 Taxonomy levels

There are three steps for a taxonomic study: (1) alpha-taxonomy (or classical taxonomy), which is concerned with the description of novel species and their classification into broad genera, (2) beta-taxonomy (or explorative taxonomy), which focuses on species-level relationships, and (3) gamma-taxonomy (or Encyclopedia taxonomy), which focuses on species-level relationships. Intraspecific differences and their evolutionary

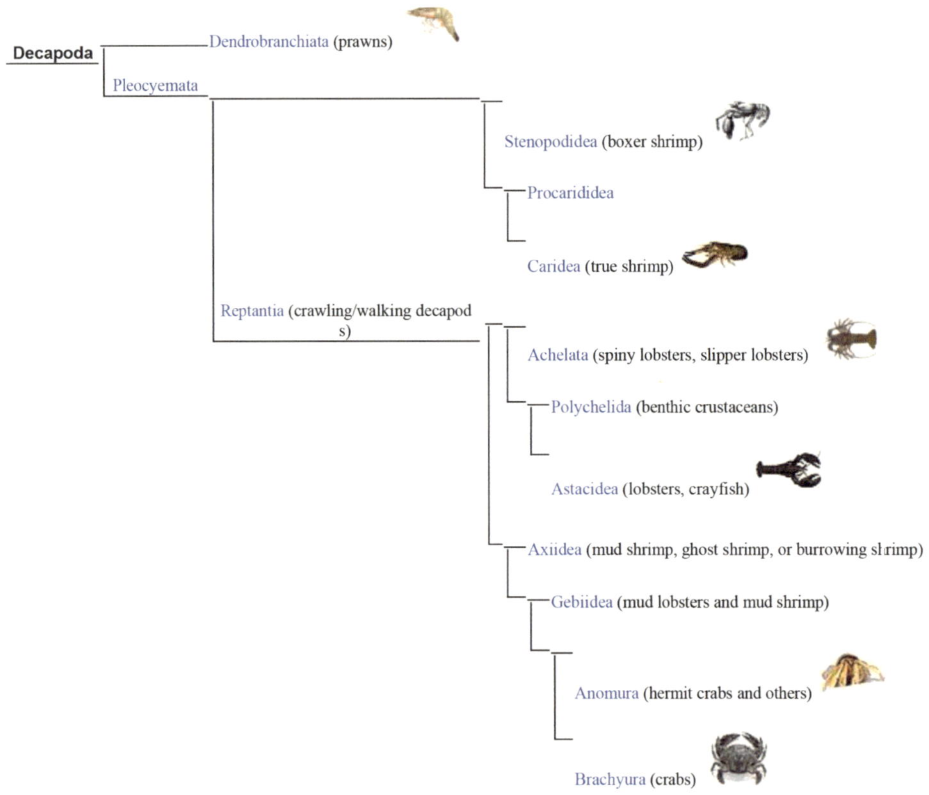

Figure 3.
Illustrate the cladogram below results from analysis by [8].

links are often known as speciation research (i.e. study of speciation). In reality, because of the three levels of taxonomy overlap, practice is impossible to investigate any species in isolation from each other taxonomy has only been applied to a few animal groups (e.g. birds and butterflies) the gamma level was reached. Study or work in the majority of animal groups including crustaceans is at gamma and alpha levels.

Linnaeus invented this system of naming, and the standards for naming animals are currently written forth in the International Code of Zoological Nomenclature (ICZN).

- Each species name in the binomial system has two parts: the genus name and the species name (also known as the specific epithet); for example, the scientific name of the kiddi (karikadi) shrimp is Parapeneopsis stylifera (H. Milne Edwards, 1837 in H. Milne Edwards 1834–1840).

- Even when derived from a proper noun such as a person's or place's name, the initial letter of the genus is always capitalized, whereas the first letter of the species is not. Every binomial scientific name is either a Latinized version of a word from another language or a Latinized version of a word from another language.

- All names of genera and lower taxa are italics (underlined while writing) by convention, while family names and higher taxa are put in plain text.

Taxon	Extant species	Extant species are also known as fossils	Exclusively fossil species
Order Decapoda Latreille, 1802	14,335	321	2979
Suborder Dendrobranchiata Bate, 1888	540	0	98
Suborder Pleocyemata Burkenroad, 1963	13,795	321	2862
Infraorder Stenopodidea Bate, 1888	69	0	2
Infraorder Caridea Dana, 1852	3268	0	57
Infraorder Astacidea Latreille, 1802	653	5	124
Infraorder Glypheidea Winkler, 1883	2	0	256
Infraorder Gebiidea de Saint Laurent, 1979	192	4	25
Infraorder Axiidea de Saint Laurent, 1979	423	14	260
Infraorder Achelata Scholtz and Richter, 1995	140	3	72
Infraorder Polychelidea Scholtz and Richter, 1995	38	0	55
Infraorde Anomura MacLeay, 1838	2451	19	230
Infraorde Brachyura Linnaeus, 1758	6559	276	1781

Table 2.
Number of species Decapoda [9].

- *Penaeus indicus* H. Milne Edwards, 1837 is a trinomial name for a subspecies of the genus Penaeus (trinomen for animals, ternary name for plants). A species' binomial name is also known as its Latin name. Biologists, on the other hand, prefer the term "scientific name" over "Latin name" because the words used to form these names are not always from the Latin language, even though terms from other languages are frequently Latinized to make them fit for this purpose. Species names are frequently drawn from Ancient Greek names [10, 11].

- Species names are frequently based on a person's surname, such as a well-known scientist's surname (e.g., *Aristeus alcocki* and *Penaeus silasi*), or a Latinized form of a relevant geographical name (e.g., *A. alcocki* and *Penaeus silasi*) (e.g., *Metapenaeus kutchensis* and *Metapeneiopsis andamanensis*) [11]. The species and generic names of binomen have opposing purposes: The species name indicates distinctness, while the generic name expresses the relationship.

- Biologists normally include the authority (Author who originally described the species) and the date (year) of the species description when using the scientific name of the species. Thus, the scientific name of the Indian shrimp species is *P. indicus* H. Milne Edwards, 1837, where the name "H. Milne Edwards" informs the reader of who originally characterized the species and 1837 is the year (year) of publication.

4. Binomial nomenclature's importance

Binomial nomenclature is widely used because it prevents the ambiguity that can arise when common names are used to refer to a species. Even within a country, common names might range from one region to the next and certainly from one country to the next. The scientific name, on the other hand, can be used anywhere in the world and in any language, avoiding confusion and translation difficulties [11]. The only internationally (universally) acknowledged standard way of referring to biological creatures is by their scientific names. They make it easier for scientists as well as trading partners who speak different languages to communicate. Without their utilization and standardization, no two individuals could truly know what organism they were discussing unless they had both seen it.

The processes used to create binomial nomenclature tend to favor consistency. Even while the existing stability is far from perfect, it is nevertheless helpful, for example, when species are transferred between genera (as is frequently the case as a result of new knowledge), the species descriptor is maintained. Similarly, historical species names may be kept as infraspecific descriptors if previously assumed to be different species relegated from species to a lower rank. Animals of subgenera and above have uninominal scientific names that begin with a capital letter. The ICZN mandates defined taxonomic ends, such as superfamily (–oidea), family (–idea), subfamily (–inae), and tribe (–inae) (–ini). The classification of shrimps exemplifies this underneath (also see **Tables 1** and **2**):

Phylum: Arthropoda incertae sedis
Subphylum: Crustacea Brünnich, 1772
Class: Malacostraca Latreille, 1802
Subclass: Eumalacostraca Grobben, 1892
Superorder: Eucarida Calman, 1904
Suborder: Dendrobranchiata Bate, 1888
Superfamily: Penaeoidea Rafinesque, 1815
Family: Penaeidae Rafinesque, 1815
Subfamily: Penaeinae Rafinesque, 1815

5. Nomenclature rules

Typification is the process of identifying a nomenclatural type. It is the method by which taxa are given scientific names. A zoological object on which the original published description of a name is based is referred to as a "type." Except for the plenary powers of ICZN, the "type" cannot be modified once it has been identified (not even by the original author). Even though Linnaeus never defined any specimen as a "type," his descriptions were based on a single specimen, and he replaced old specimens with fresh ones. This method persisted throughout Europe for a long time, causing uncertainty when it came to tracing the original specimens. The entomologist Pierre Andre Latreille, who lived in the early 1800s, is thought to have started the practice of specifically defining types [10, 12]. Not all early authors kept type material, which can make later identification of species more difficult.

In 1901, the zoological code mandated typification for future work. There are 41 various type series; however, the following are the most well known [11]:

1. **Holotype**: A single specimen chosen by the author of a species to serve as the genuine type for the species. A holotype must be labeled with the following information: location, date, size, sex, developmental stage, the hostname (if epizoic or parasitic), and collector name. It should be deposited in a reputable museum and given a unique identification number.

2. **Paratype**: A specimen other than the holotype cited in the original description.

3. **Allotype**: An allotype is a specimen of the opposite gender as the type.

4. **Syntype**: Specimens used to determine the principal type (holotype or neotype).

5. **Topotypes**: Specimens that have been identified as having a unique origin.

6. **Apotype**: A specimen rather than the type on which a subsequent or supplemental description or illustration is based. It is also known as a hypotype.

7. **Genotype**: A species that is defined as the type species of the genus it is based on.

8. **Lectotype**: The type of a species is a specimen chosen from a group of syntypes after the description has been published.

9. **Neotype**: When a species' designated type (holotype, syntypes, neotype, or lectotype) is previously designated but then lost or damaged beyond recognition, a neotype is a new specimen recognized as the type.

10. **Isotype:** A specimen acquired at the same time as the holotype, paratypes, or syntypes, but not from the same location [11, 13].

11. **Monotype**: A monotype occurs when the holotype of a species is the only specimen known to science [13].

12. **Morphotype**: A specimen depicting a species' morphological variation [11, 13].

6. Synonymy

Synonymy is a term used to describe two or more names that belong to the same taxon. The Law of Priority states that when a species has multiple synonyms, only the oldest one is valid (ICZN) [11]. The oldest is treated as the proper name and is referred to as a senior synonym, whereas the others are referred to as junior synonyms. Synonyms are a major tricky for taxonomists. These are made owing to a lack of awareness of the available research or the amount of variation that a species can have. Many times, the same species is reported by two or more writers under multiple names without the authors being aware of the species differences (**Table 3**).

Order	Suborder	Superfamily	Family	Genus
			Hollinidae Swartz, 1936	*Anticostiella* Copeland, 1973
			Tetradellidae Swartz, 1936	*Foramenella* Stumbur, 1956
				Tetradella Ulrich, 1890
	Palaeocopina Henningsmoen, 1953	Hollinoidea Swartz, 1936	Oepikellidae Jaanusson, 1957	*Platybolbina* Henningsmoen, 1953
				Eographiodactylus Kraft, 1962
			Tvaerenellidae Jaanusson, 1957	*Tvaerenella* Jaanusson, 1957
		Eurychilinoidea Ulrich and Bassler, 1923	Eurychilinidae Ulrich and Bassler, 1923	*Eurychilina* Ulrich, 1889
			Bolliidae Bouček, 1936	*Jonesites* Coryell, 1930
				Dicranella Ulrich, 1894
Beyrichiocopida Pokorný, 1954			Circulinidae Neckaja, 1966	*Byrsolopsina* Swain and Cornell, 1961
				Easchmidtella Schallreuter, 1967
	Binodicopina Schallreuter, 1972	Binodicopina Schallreuter, 1972	Aechminidae Swartz, 1936	*Aechmina* Jones and Holl, 1869
				Pseudohippula Schallreuter, 1975
			Spinigeritidae Schallreuter, 1980	*Faurella* Copeland, 1973
				Spinigerites Schallreuter, 1980
			Schmidtellidae Neckaja, 1966	*Schmidtella* Ulrich, 1892
	Paraparchitocopa Gramm, 1975		Jaanussoniidae Schallreuter, 1971	*Hemiaechminoides* Morris and Hill, 1952
				Hemeaschmidtella Schallreuter, 1971
	Leiocopina Schallreuter, 1973b	Aparchitoidea Jones (*in* Chapman), 1901	Aparchitidae Jones (*in* Chapman), 1901	*Baltonotella* Sarv, 1959
	Eridostraca Adamczak, 1961		Eridoconchidae Henningsmoen, 1953	*Cryptophyllus* Levinson, 1951
			Steusloffinidae Schallreuter, 1984	*Medianella* Neckaja, 1966

Order	Suborder	Superfamily	Family	Genus
				Longiscula Neckaja, 1958
			Longisculidae Neckaja, 1966	*Aviacypris* Schallreuter, 1977
Podocopida Sars, 1866	Metacopina Sylvester-Bradley, 1961		Krausellidae Berdan, 1961	*Krausella* Ulrich, 1894
			Rectellidae Neckaja, 1966	*Rectella* Neckaja, 1958
		Bairdiocypridoidea Shaver, 1961	Bairdiocyprididae Shaver, 1981	*Elliptocyprites* Swain, 1962
			Pachydomellidae Berdan and Sohn, 1961	*Microcheilinella* Geis, 1933
		Thlipsuroidea Ulrich, 1894	Thlipsuriidae Ulrich, 1894	*Anticostus* n. gen
	Cypridocopina Jones, 1901	Bairdioidea Sars, 1865	Macrocyprididae Muller, 1912	*Macrocyproides* Spivey, 1939
	Cytherelliformes Skogsberg, 1920	Leperditelloidea Ulrich and Bassler, 1906	Leperditellidae Ulrich and Bassler, 1906	*Leperditella* Ulrich, 1894
Platycopida Sars, 1866	Kloedenellocopina Scott, 1961	Kloedenelloidea Ulrich and Bassler, 1908	Kloedenellidae Ulrich and Bassler, 1908	*Eokloedenella* Kraft, 1962

Table 3.
Taxonomic structure of late Ordovician ostracod [5].

7. Priority

This is a contentious aspect of zoological nomenclature, yet it is a principal ICZN rule that promotes stability. When two names for the same taxon are identified, the law of priority determines which one is valid. With few exceptions, the valid name is the oldest (as determined by published sources). The authority of a name in a family, genus, or species does not alter if its rank within the group rises or falls. According to Ref. [14] nomenclature of the Pentastomida with a list of species and used higher taxon, Pentastomida is a name that can be traced back to at least Huxley (1869). General taxonomical structure of the late Ordovician ostracod fauna of the Ellis Bay Formation, Anticosti Island, eastern Canada (see **Table 1**).

8. Classification of Crustacean Decapoda

Classification for the entire Crustacean Decapoda has been updated. There are 2725 genera and 17,635 species of decapods in the 233 families that make up the categorization (including both extant and fossil species **Figure 4**). The families in our taxonomy consist of 71 current species only, 109 fossil species, and 53 entirely fossil species. A total 14,756 species are now thought to be extant, while 2979 species are only known from fossils [11].

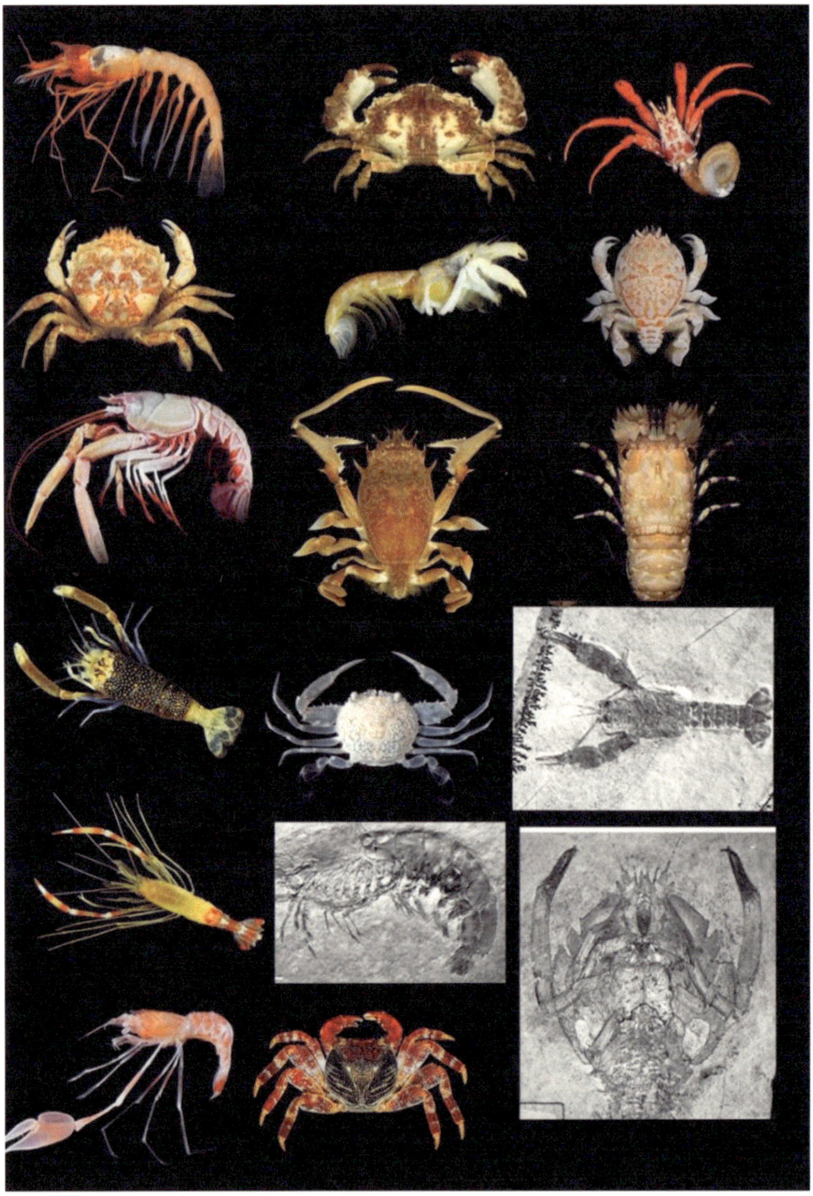

Figure 4.
Genera of Decapoda Crustaceans, both modern and extinct [10].

Author details

Zardasht Ahmed Taha
Department of Geology, College of Science, University of Sulaimani, Iraq

*Address all correspondence to: zardasht.taha@univsul.edu.iq

References

[1] Bowman TE, Abele LG. Classification of the recent Crustacean. In: Abele LG, Bliss dE, editors. Systematics, the Fossil Record, and Biogeography, the Biology of Crustacean. Vol. 1. New York: Academic Press; 1982. pp. 1-27

[2] Hobbs H, Crustacea H. Encyclopedia of Caves and Karst Science. London: Routledge; Retrieved December 5, 2006

[3] Martin JW, George E. Davis. An updated classification of the recent Crustacea. Vol. 39. Los Angeles: Natural History Museum of Los Angeles County. 2001

[4] Simpson GG. Principles of animal taxonomy. In Principles of Animal Taxonomy. Columbia University Press. 1961

[5] Taha ZA. The taxonomic, biogeographical, palaeogeographical, and palaeoecological significance of the Late Ordovician ostracod Fauna of the Ellis Bay formation, Anticosti Island, eastern Canada. A thesis submitted for the degree of Doctor of Philosophy at the University of Leicester. 2018. 182

[6] Taha ZA. Late Cretaceous Ostracoda in the Shiranish Formation Dokan Area, Kurdistan Region-NE Iraq. 2020

[7] Ghafoor IM. Crustacean. Crustacea. London, UK, London: IntechOpen; 2020

[8] Wolfe JM, Breinholt JW, Crandall KA, Lemmon AR, Lemmon EM, Timm LE, et al. A phylogenomic framework, evolutionary timeline and genomic resources for comparative studies of decapod crustaceans. Proceedings of the Royal Society B. 2019:286

[9] De Grave S, Pentcheff ND, Ahyong ST. A classification of living and fossil genera of decapod crustaceans. Raffles Bulletin of Zoology. 2009

[10] De Grave S, Pentcheff ND, Ahyong ST, Chan TY, Crandall KA, Dworschak PC, et al. A classification of living and fossil genera of decapod crustaceans. Raffles Bulletin of Zoology. Supplement No. 21: 2009:1-109

[11] Josileen J, Pillai SL. Training Programme on taxonomy and identification of commercially important crustaceans of India. 2013

[12] Dubois C. Pierre Andre Latreille (1762-1833): The foremost entomologist of his time. Annual Review of Entomology. 1974;**19**:1-14

[13] Coad BW, McAllister DE. Dictionary of ichthyology. 2006. Available from: http://www.coad.ca/Dictionary/CompleteDictionarylatestversion.htm

[14] Poore GCB. The nomenclature of the recent Pentastomida (Crustacea), with a list of species and available names. Systematic Parasitology. 2012;**82**:211-240

Chapter 2

Eco-Morphology of Some Decapod Crustaceans in a Tropical Coastal Marine Waters

Adefemi O. Ajibare, Olaronke O. Olawusi-Peters and Joshua O. Akinola

Abstract

The relationship among the morphology, population of crustaceans and water quality of the coastal marine waters of Ondo State, Nigeria, was assessed in order to accentuate the sustainability of biodiversity in the coasts. Standard methods were employed to identify and examine the effect of the environment on the crustaceans. The DO (7.58 mg/l), temperature (29.53°C), pH (6.69), turbidity (44.03NTU), salinity (16.48‰), hardness (85.88 mg/l), biochemical oxygen demand (21.22 mg/l) and conductivity (41.55 µS cm^{-1}). The population structure of decapod crustaceans follows the order *Nematopalaemon hastatus* > *Farfantepenaeus notialis* > *Holthuispenaeopsis atlantica* > *Macrobrachium macrobrachion* > *Sanquerus validus* > *Ocypode africana* > *Callinectes marginatus*. The sampled organisms (*F. notialis*, *M. macrobrachion*, *N. hastatus* and *Holthuispenaeopsis atlantica*) had mean total length (cm) (9.41 ± 1.62, 7.14 ± 0.77, 6.69 ± 0.81 and 11.78 ± 0.60) and body weight (g) of (3.21 ± 1.63, 2.37 ± 0.79, 1.34 ± 0.56 and 6.72 ± 0.47 g), respectively. *C. marginatus*, *Ocypode africana* and *Sanquerus validus* had a mean carapace length (cm) of 5.50 ± 0.71, 4.83 ± 1.27 and 8.31 ± 3.50, respectively, and mean body weight (g) of 4.69 ± 0.95, 3.41 ± 4.72 and 66.21 ± 50.45, respectively. PCA revealed strong correlation among BOD, DO and the morphological parameters of each species. Also, Single Factor and Comprehensive Pollution Indices revealed a slightly and moderately polluted aquatic ecosystem, respectively. Thus, adequate control of all activities in the ecosystem for healthy growth and survival of aquatic species is essential.

Keywords: shrimps, crabs, population, pollution, principal components analysis

1. Introduction

Crustaceans are among the most diverse, numerous, and widely distributed decapods. They are found all across the tropics and are key commercial fisheries for a country's economy. The majority of the species are only found in estuaries, and many of them require brackish or coastal water for larval development [1].

The relationships between the forms and functions of organisms, as well as the ecological qualities connected with the utilization of acquired resources, are studied in eco-morphological studies [2, 3]. The research answers fundamental concerns about organism niches, shared resources, and community organization, as well as basic techniques to developing a "fit" between organisms and their environments. Climate change, habitat destruction, and other forms of pollution are all contributing to the loss of marine biodiversity around the world. Because to changing environmental variables, fish have faced a greater vulnerability threat. Ecomorphological research has been increasingly important in recent years, particularly as the environment's impact on organism growth and survival has increased. Among vertebrates, aquatic species have the most morphological variety and exhibit a wide and strong link between form and function [4]. Thus, they can serve as a vital tool in understanding and establishing relationships between ecology and morphology.

Coastal waters are essential to life and concentrate more than half of the world's population of aquatic life. Although the areas have long been considered an unlimited resource, the health and biodiversity of the resource of the zones are being threatened with some fish stocks collapsing due to uneven environmental factors, changing chemistry, over-exploitation, pollution etc [5]. These have a profound impact on reproduction and survival of various species, ecological adaptation and community and productivity of the ecosystem [6].

The coastal marine waters of Nigeria have been adduced to be areas of high biodiversity of aquatic species with the composition, distribution, abundance and growth widely linked to numerous ecological factors. Several authors [7–12] have worked on decapods in the coastal marine waters of Ondo State, Nigeria and reported the overall wellbeing of the species as well as the level of pollution in the environment. The authors strongly suggested the need to examine the eco-morphology of decapods in the coastal environment in order to determine the effecsof the environment on the growth and well-being of the crustaceans. Recommendations from the various authors and dearth of information on ecomorphology of widely distributed decapods in coastal marine waters of Ondo State, Nigeria necessitated this study.

Since the organization of any community (atmospheric, aquatic, terrestrial) can best be understood through the relationship of organisms and how they acquire, use and share resources with each other, it is therefore essential to assess and study the eco-morphology of decapods in the coastal marine waters of Ondo State, Nigeria which is known for a diverse assemblage of fish species to understand the conditions of the ecosystem.

2. Materials and methods

2.1 Study area

The study was carried out in Ayetoro (Station I), Idiogba (Station II), Asumogha (Station III) and Bijimi (Station IV) of Ilaje Local Government Area of Ondo State from September to December 2011. The study area falls within Latitude 5° 50′N–6° 09′N and Longitude 4° 45′E – 5° 05′E of the Greenwich Meridian. The stations were selected based on extensive shrimp fishing in the communities.

2.2 Sample collection and identification

Shrimps were collected monthly with the assistance of artisanal fishermen and were immediately preserved in ice chest before being transferred to the Fisheries and Aquaculture Laboratory of the Federal University of Technology, Akure where identification was done according to FAO Species Identification Sheets, (Volume VI) [13].

2.3 Morphometric measurement

Measurements were taken for total length (tip of rostrum to edge of telson), carapace length (posterior margin of carapace to edge of telson), rostral length (tip of rostrum to the posterior end of the orbit), and body length (posterior margin of the orbit to the edge of telson) to the nearest 0.01 cm using graduated measuring board while weight was determined with top loading digital Sartorius weighing balance (Model 1100) to the nearest 0.01 g.

2.4 Water sampling and analysis

Water samples from each station were collected monthly at sub-surface level with 250 ml sampling bottles and transported in ice chest to the Fisheries and Aquaculture Laboratory of the Federal University of Technology, Akure for analysis. Samples handling and preservation were done following the standard method [14]. The temperature, turbidity and conductivity of the water were done in-situ with a standard mercury-in-glass thermometer, turbidity meter and Knick Portamess conductivity Meter (Model 913) respectively while pH, Salinity and Dissolved Oxygen (DO) were determined using the Hanna multi-parameter kit (Model H19828).

2.5 Data analysis

Data obtained from physical and chemical measurements were subjected to multivariate analysis of variance (MANOVA) using the Statistical Package for Social Sciences (SPSS), Version 20.0 and was tested at $P < 0.05$ for significance. The mean values were compared with the water quality criteria of World Health Organization (WHO) and Nigerian Federal Environmental Protection Agency (FEPA). Morphology and water parameter data were related with Principal Components Analysis (PCA) while the water pollution status was determined using both single-factor pollution index (SFPI) andcomprehensive pollution index (CPI).

The single-factor pollution index is defined as [15]:

$$P_i = \frac{C_i}{S_i} \tag{1}$$

Where: P_i is the pollution index of pollutant i,

C_i is the measured concentration of the pollution indicator (mg/l).

S_i is the National water quality standard permissible limit for the pollution indicator in surface water.

The water quality factor P_i is classified into five grades, as listed in **Table 1** [16].

P_i	Pollution grades
Less than 0.4	Non-pollution
0.4–1.0	Slight pollution
1.0–2.0	Medium polluted
2.0–5.0	Heavy polluted
More than 5.0	Serious polluted

Table 1.
Standard grades for single-factor pollution index (PI).

The comprehensive pollution index (CPI)	Level	Explain the water quality grades
Less than 0.2	I	Cleanness
0.21–0.4	II	Sub-cleanness
0.41–1.0	III	Slight pollution
1.01–2.0	IV	Moderate pollution
More than 2.01	V	Severe pollution

Table 2.
Standard surface water quality categories based on CPI.

The **comprehensive pollution index (CPI)** is defined as follows [17]:

$$\mathrm{CPI} = \frac{1}{n}\sum_{i=1}^{n}\frac{C_i}{S_i} \tag{2}$$

Where: CPI = the comprehensive pollution index,

C_i = the measured concentration of the pollution indicator (mg/l),

S_i = National water quality standard permissible limit for the pollution indicator in surface water.

n = the number of chosen parameters.

CPI is classified into the water quality levels listed in **Table 2** according to [17].

3. Results

3.1 Population structure of sampled organisms

The population structure of the sampled organisms is presented in **Table 3**. The table revealed a total of 225 *Farfantepenaeus notialis*, 27 *Macrobrachium macrobrachion*, 1999 *Nematopalaemon hastatus*, 134 *Holthuispenaeopsis atlantica*, 2 *Callinectes marginatus*, 3 *Ocypode africana* and 18 *Sanquerus validus*. The most predominant species was *N. hastatus*. *M. macrobrachion* was not recorded in Station III, while *C. marginatus* and *Ocypode africana* were only recorded in Station III.

Species	Station I	Station II	Station III	Station IV	Total
Farfantepenaeus notialis	51	73	53	48	225
Macrobrachium macrobrachion	13	4	0	10	27
Nematopalaemon hastatus	526	601	416	456	1999
Holthuispenaeopsis atlantica	24	29	20	61	134
Callinectes marginatus	0	0	2	0	2
Ocypode africana	0	0	3	0	3
Sanquerus validus	5	2	6	5	18

Table 3.
Population structure of sampled organisms.

3.2 Morphological parameters of sampled organisms

The morphological features of the sampled organisms are presented in **Tables 4** and **5**. The morphometric measurements of sampled shrimps varied across the sampled species. *F. notialis*, *M. macrobrachion*, *N. hastatus*and *Holthuispenaeopsis atlantica* had mean total length (cm) of 9.41 ± 1.62, 7.14 ± 0.77, 6.69 ± 0.81 and 11.78 ± 0.60 respectively and mean rostral length (cm) of 2.72 ± 0.47, 2.48 ± 0.25, 2.31 ± 0.35 and 2.78 ± 0.17 respectively. The mean carapace length (cm) of *F. notialis*, *M. macrobrachion*, *N. hastatus* and *H. atlantica* obtained in the aquatic ecosystem was 2.70 ± 0.44, 2.36 ± 0.25, 2.20 ± 0.40 and 2.80 ± 0.22 cm respectively while the corresponding mean body length (cm) was 5.63 ± 0.92, 4.15 ± 0.53, 4.49 ± 0.54 and 6.30 ± 0.51respectively. The body weight (g) of *F. notialis*, *M. macrobrachion*, *N. hastatus*and *H. atlantica* obtained in the aquatic ecosystem was 3.21 ± 1.63, 2.37 ± 0.79, 1.34 ± 0.56 and 6.72 ± 0.47 respectively (**Table 4**). The table further shows that *N. hastatus* had varying ranges of morphological features across the stations.

The morphometric measurements of sampled crabs which also varied across the sampled species are presented in **Table 5**. *C. marginatus*, *O. africana* and *S. validus* had a mean carapace length (cm) of 5.50 ± 0.71, 4.83 ± 1.27 and 8.31 ± 3.50 respectively, and mean body weight (g) of 4.69 ± 0.95, 3.41 ± 4.72 and 66.21 ± 50.45 respectively (**Table 5**).

3.3 Water quality parameter of the coastal marine waters of Ondo State Nigeria

The water quality parameters obtained in the coastal marine waters of Ondo state is presented in **Table 6**.

The Table shows that the parameters excluding turbidity and conductivity had no significant differences across the stations. The mean DO concentration was 7.58 mg/l, while water temperature and pH had a mean concentration of 29.53°C and 6.69 respectively. The turbidity and salinity of the aquatic ecosystem had a mean concentration of 44.03NTU and 16.48‰ respectively. Hardness, biochemical oxygen demand (BOD) and conductivity of the water body recorded mean values of 85.88 mg/l, 21.22 mg/l and 41.55 μS cm^{-1} respectively.

Parameter	Species	Station I	Station II	Station III	Station IV	Mean
Total length (cm)	*Farfantepenaeusnotialis*	9.12 ± 1.74	9.34 ± 1.68	9.67 ± 1.44	9.56 ± 1.57	9.41 ± 1.62
	Macrobrachium macrobrachion	7.40 ± 0.80	6.70 ± 0.39	—	6.98 ± 0.77	7.14 ± 0.77
	Nematopalaemon hastatus	6.70 ± 0.45	7.00 ± 0.13	6.09 ± 1.15	6.85 ± 0.97	6.69 ± 0.81
	Holthuispenaeopsisatlantica	11.51 ± 1.87	11.94 ± 1.77	12.80 ± 1.70	11.04 ± 2.53	11.78 ± 0.60
Rostral length (cm)	*Farfantepenaeus notialis*	2.63 ± 0.37	2.75 ± 0.49	2.78 ± 0.50	2.69 ± 0.48	2.72 ± 0.47
	Macrobrachium macrobrachion	2.46 ± 0.17	2.50 ± 0.08	—	2.49 ± 0.36	2.48 ± 0.25
	Nematopalaemon hastatus	2.50 ± 0.20	2.53 ± 0.09	1.98 ± 0.36	2.19 ± 0.38	2.31 ± 0.35
	Holthuispenaeopsisatlantica	3.18 ± 0.40	3.38 ± 0.48	3.35 ± 0.50	3.10 ± 0.67	2.78 ± 0.17
Carapace length (cm)	*Farfantepenaeus notialis*	2.67 ± 0.39	2.66 ± 0.46	2.83 ± 0.47	2.68 ± 0.41	2.70 ± 0.44
	M. macrobrachion	2.38 ± 0.22	2.50 ± 0.00	—	2.29 ± 0.32	2.36 ± 0.25
	N. hastatus	2.38 ± 0.26	2.51 ± 0.03	1.81 ± 0.42	2.05 ± 0.40	2.20 ± 0.40
	Holthuispenaeopsis atlantica	3.18 ± 0.42	3.22 ± 0.59	3.35 ± 0.57	3.04 ± 0.65	2.80 ± 0.22
Body Length (cm)	*F. notialis*	5.32 ± 1.06	5.51 ± 0.86	5.88 ± 0.89	5.85 ± 0.77	5.63 ± 0.92
	M. macrobrachion	4.20 ± 0.48	4.60 ± 0.39	—	3.90 ± 0.54	4.15 ± 0.53
	N. hastatus	4.60 ± 0.40	4.86 ± 0.25	3.98 ± 0.62	4.45 ± 0.51	4.49 ± 0.54
	Holthuispenaeopsis atlantica	6.78 ± 0.89	7.03 ± 1.15	7.26 ± 0.92	6.68 ± 1.21	6.30 ± 0.51
Weight (g)	*F. notialis*	3.39 ± 2.14	2.88 ± 1.33	3.41 ± 1.58	3.29 ± 1.46	3.21 ± 1.63
	M. macrobrachion	2.67 ± 0.77	2.47 ± 0.29	—	1.94 ± 0.80	2.37 ± 0.79
	N. hastatus	1.11 ± 0.37	1.34 ± 0.53	1.51 ± 0.71	1.43 ± 0.60	1.34 ± 0.56
	Holthuispenaeopsis atlantica	6.24 ± 4.14	7.22 ± 3.33	8.34 ± 4.45	6.24 ± 3.68	6.72 ± 0.47

Table 4.
Morphological parameters of sampled shrimps.

Parameter	Species	Station I	Station II	Station III	Station IV	Mean
Carapace Length (cm)	*Callinectes marginatus*	—	—	5.50 ± 0.71	—	5.50 ± 0.71
	Ocypode africana	—	—	4.83 ± 1.27	—	4.83 ± 1.27
	Sanquerus validus	7.96 ± 3.46	8.05 ± 2.33	8.97 ± 4.23	7.96 ± 3.92	8.31 ± 3.50
Weight (g)	*Callinectes marginatus*	—	—	4.69 ± 0.95	—	4.69 ± 0.95
	Ocypode africana	—	—	3.41 ± 4.72	—	3.41 ± 4.72
	Sanquerus validus	53.17 ± 45.66	31.65 ± 3.07	88.48 ± 58.89	66.37 ± 53.28	66.21 ± 50.45

Table 5.
Morphological parameters of sampled crabs.

Station	Station I	Station II	Station III	Station IV	Mean	FEPA [18], FEM [19]
DO (mg/l)	7.66 ± 0.22^a	7.54 ± 0.14^a	7.60 ± 0.20^a	7.53 ± 0.15^a	7.58 ± 0.18	5.0–9.0
Temperature (°C)	29.50 ± 0.90^a	29.50 ± 0.52^a	29.71 ± 0.45^a	29.42 ± 0.51^a	29.53 ± 0.61	28.0–30.0
pH	6.68 ± 0.11^a	6.71 ± 0.13^a	6.66 ± 0.05^a	6.69 ± 0.14^a	6.69 ± 0.11	6.5–8.5
Turbidity (NTU)	44.87 ± 4.97^b	43.94 ± 2.86^{ab}	45.36 ± 2.18^b	41.95 ± 4.42^a	44.03.89	≤50
Salinity (ppt)	16.55 ± 5.16^a	16.65 ± 5.36^a	16.35 ± 4.95^a	16.36 ± 5.11^a	16.48 ± 4.98	1.0–15
Hardness (mg/l)	87.16 ± 3.47^a	84.75 ± 1.84^a	84.57 ± 1.80^a	87.03 ± 5.13^a	85.88 ± 3.47	50–100
BOD (mg/l)	22.10 ± 4.96^a	22.50 ± 3.90^a	20.74 ± 6.45^a	19.55 ± 2.62^a	21.22 ± 4.70	3.0–6.0
Conductivity (μS cm^{-1})	41.84 ± 18.95^b	41.10 ± 18.81^a	41.60 ± 12.97^b	41.67 ± 17.75^b	41.55 ± 16.97	<35,000

Table 6.
Water quality parameter of the coastal marine waters of Ondo state Nigeria.

3.4 Water pollution assessment

The Single-Factor Pollution Index (PI) and Comprehensive Pollution Index (CPI) of the Coastal Marine Waters are presented in **Table 7**. The Table shows that the values obtained for DO (1.26), Conductivity (1.19) and pH (1.01) classified the study area to be moderately polluted as the mean values were within 1.0–2.0 (as earlier stated in **Table 1**), while Temperature (0.98), Turbidity (0.88), Salinity (0.47) and Hardness (0.86) classified the water as slightly polluted (with values within 0.40–1.0 as interpreted in **Table 1**). The mean BOD (4.24) indicated heavy pollution as the value was within 2.0–5.0 (as in **Table 1**). The mean CPI (1.36) showed moderate pollution as the value was within 1.0–2.0 of the standard surface water quality (**Table 2**).

3.5 Relationship between the morphology of sampled organisms and water quality

The Principal Components Analysis of the morphological parameters of sampled organisms and water quality parameters is presented in **Table 8** (Shrimps) and 9 (Crabs) as well as **Figures 1–7**. The results generally revealed that there was a strong

Index	Parameter	Station I	Station II	Station III	Station IV	Mean
Single-Factor Pollution Index (SPI)	DO	1.28	1.26	1.27	1.25	1.26
	Temperature	0.98	0.98	0.99	0.98	0.98
	pH	1.01	1.02	1.01	1.01	1.01
	Turbidity	0.9	0.88	0.91	0.84	0.88
	Salinity	0.47	0.48	0.47	0.47	0.47
	Hardness	0.87	0.85	0.85	0.87	0.86
	BOD	4.42	4.5	4.15	3.91	4.24
	Conductivity	1.2	1.17	1.19	1.19	1.19
Comprehensive Pollution Index (CPI)		1.39	1.39	1.35	1.32	1.36

Table 7.
Single-factor pollution index (PI) and comprehensive pollution index (CPI) of the coastal marine waters of Ondo state Nigeria.

correlation within the morphological parameters of each species of shrimps and crabs. Also, the important pollution indicator water quality parameters (Temperature, DO and BOD), all loaded positively (in the same principal component) with the morphological characteristic of each examined species (**Tables 8** and **9**). **Figures 1**–**7** further buttressed that parameters (principal components) in the same circle showed a positive correlation with one another. Similarly, pH, salinity, conductivity and turbidity loaded positively in the same principal component but negatively with temperature, DO and BOD (**Tables 8** and **9**). This shows that the parameters contributed significantly to the survival, growth and abundance of each species of shrimps and crabs in the water body.

	Farfantepenaeus notialis		***Macrobrachium macrobrachion***		***Nematopalaemon hastatus***		***Holthuispenaeopsis atlantica***	
	PC1	**PC2**	**PC1**	**PC2**	**PC1**	**PC2**	**PC1**	**PC2**
Total length(cm)	0.97			0.88	0.87		0.96	
Rostral length (cm)	0.87			0.70	0.93		0.91	
Carapace length (cm)	0.90			0.70	0.88		0.94	
Body length (cm)	0.92		−0.21	0.78	0.92		0.96	
Weight (g)	0.88		−0.13	0.80	0.35	−0.14	0.93	
DO	0.27		−0.16	0.13			0.19	−0.10
Temperature	0.77		0.33	0.34	0.13		0.11	0.17
pH		0.69	0.71		0.31	0.60		0.68
Turbidity		0.57	0.55		0.21	0.78		0.76
Salinity		0.93	0.95	0.16	−0.13	0.94		0.95
Hardness		0.72	0.69	−0.18	−0.24	0.59	−0.14	0.68
BOD	0.14	−0.83	−0.86		0.37	−0.70	0.11	−0.80
Conductivity		0.91	0.90	0.14	−0.17	0.89		0.93

Table 8.
Relationships (principal components analysis) of the morphology of shrimps and water quality.

	Ocypode africana		*Sanquerus validus*		*Callinectes marginatus*	
	PC1	**PC2**	**PC1**	**PC2**	**PC1**	**PC2**
Carapace length (cm)	1.00		−0.10	0.43	−0.21	0.98
Weight (g)	1.00		−0.06	0.41	0.16	0.99
DO	−0.15	0.99		−0.01		1.00
Temperature	0.91	−0.42		0.03	0.32	−0.95
pH	0.59	0.81	0.15	0.17	0.87	0.49
Turbidity	0.64	0.77	0.08	0.24	−1.00	
Salinity	1.00		0.23	0.01	0.98	−0.20
Hardness	0.51	−0.86	0.21	−0.04	1.00	
BOD	−0.89	0.46	−0.24	0.05	−0.98	0.18
Conductivity	0.99	0.12	0.25	−0.06	0.48	0.88

Table 9.
Relationships (principal components analysis) of the morphology of crab and water quality.

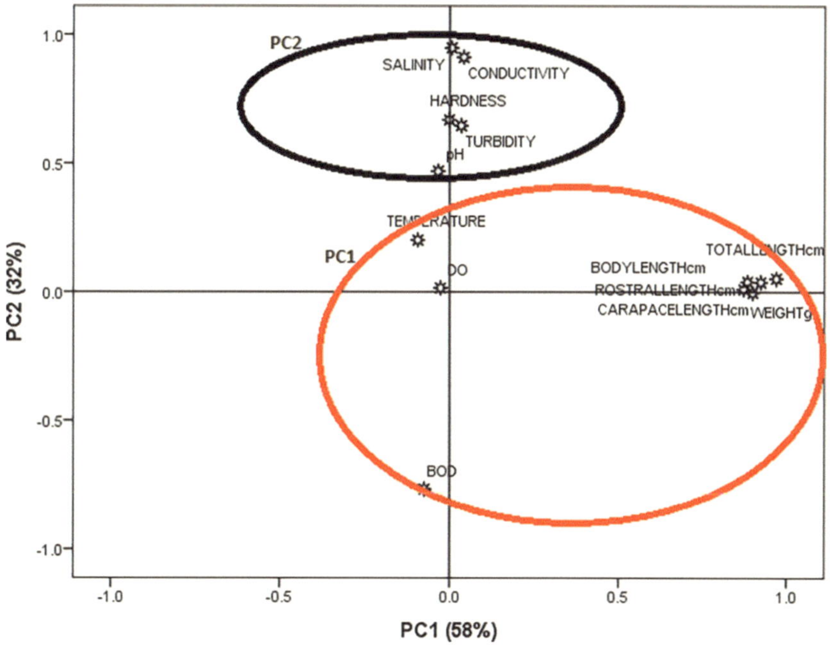

Figure 1.
Relationships (PCA) of the morphology of Farfantepenaeus notialis *and water quality.*

4. Discussion

The physical and chemical parameters of an aquatic ecosystem determine the quality of biodiversity in the environment, overall health and condition of the habitat. The parameters examined in the study area were within the tolerable range for coastal marine waters. The DO concentration was within the 4.5–8.5 mg/l recommended for the growth and survival of aquatic species within the ecosystem [18], while the

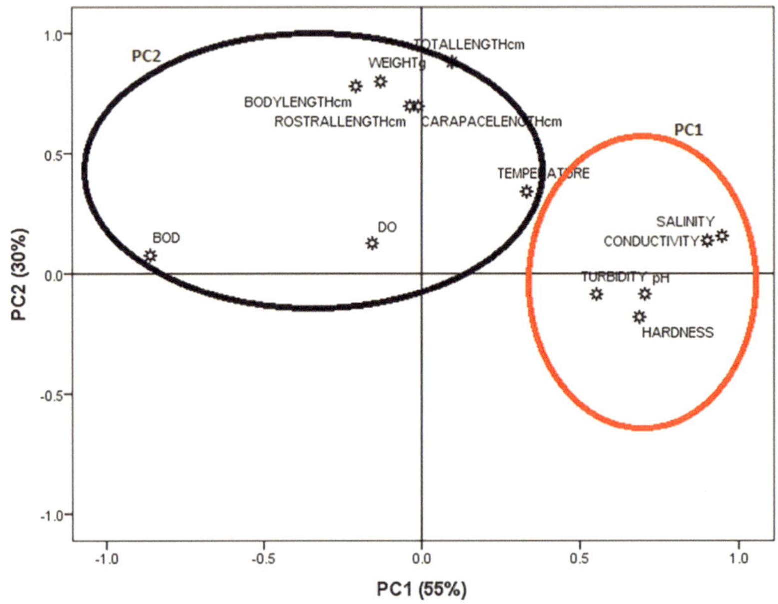

Figure 2.
Relationships (PCA) of the morphology of Macrobrachium macrobrachion *and water quality.*

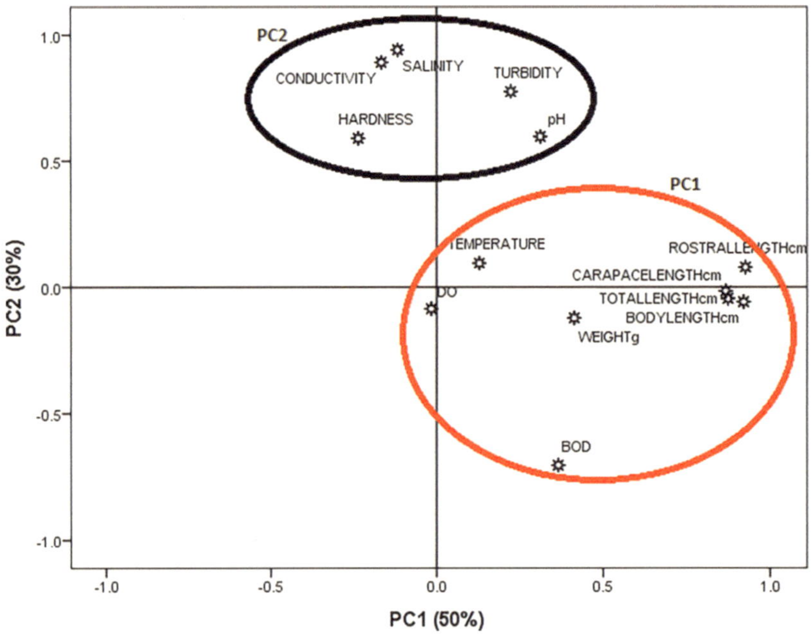

Figure 3.
Relationships (PCA) of the morphology of Nematopalaemon hastatus *and water quality.*

slightly acidic water pH was similar to the observations of Deekae et al. [20] and Bolarinwa et al. [21]. The high DO concentration and slightly acidic pH value could result from photosynthesis by large amount of plants. Great uncontrolled plant growth, especially algal blooms increases the nutrient levels in the ecosystem, thus

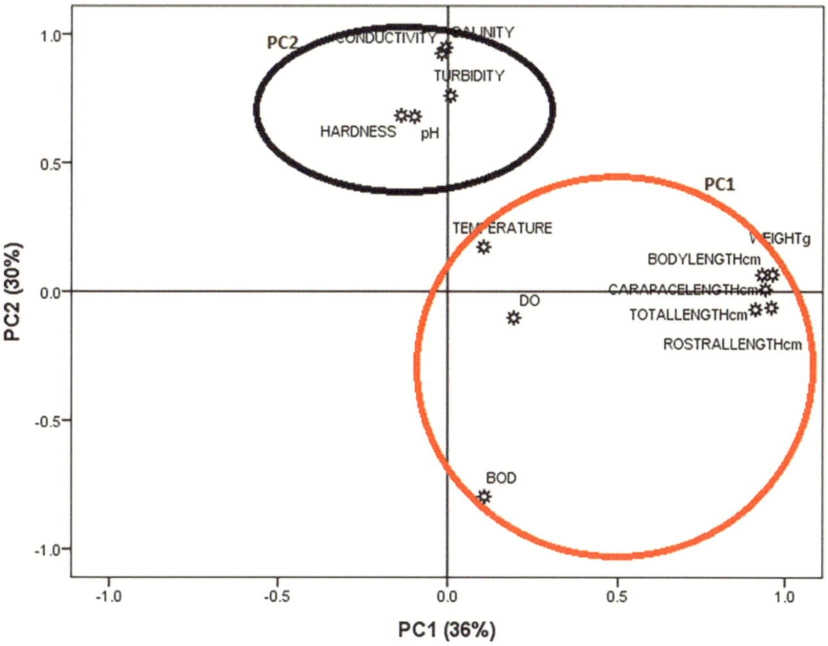

Figure 4.
Relationships (PCA) of the morphology of Holthuispenaeopsis atlantica *and water quality.*

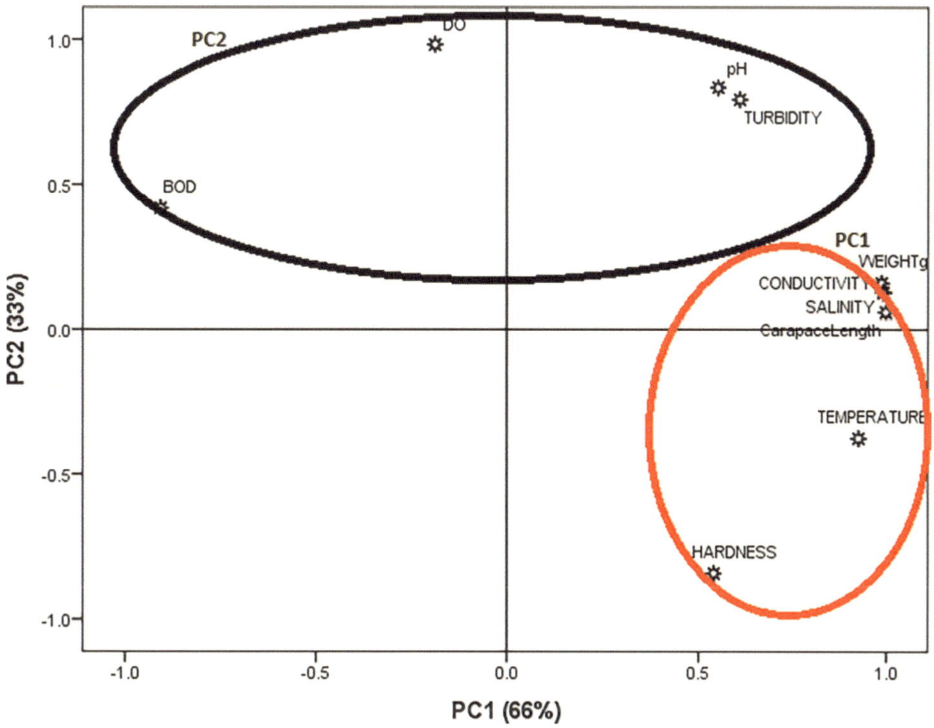

Figure 5.
Relationships (PCA) of the morphology of Ocypode africana *and water quality.*

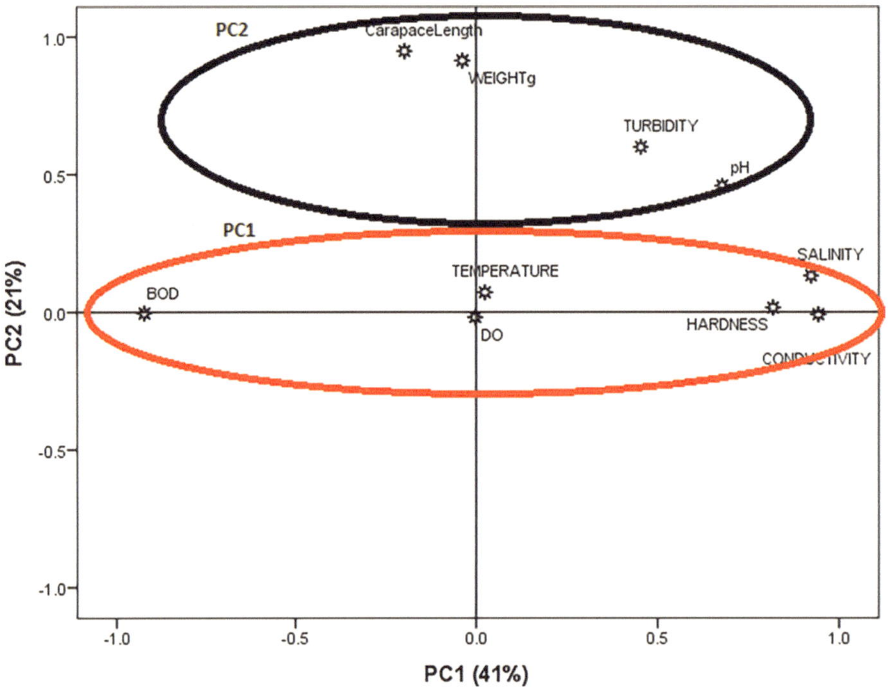

Figure 6.
Relationships (PCA) of the morphology of Sanquerus validus *and water quality.*

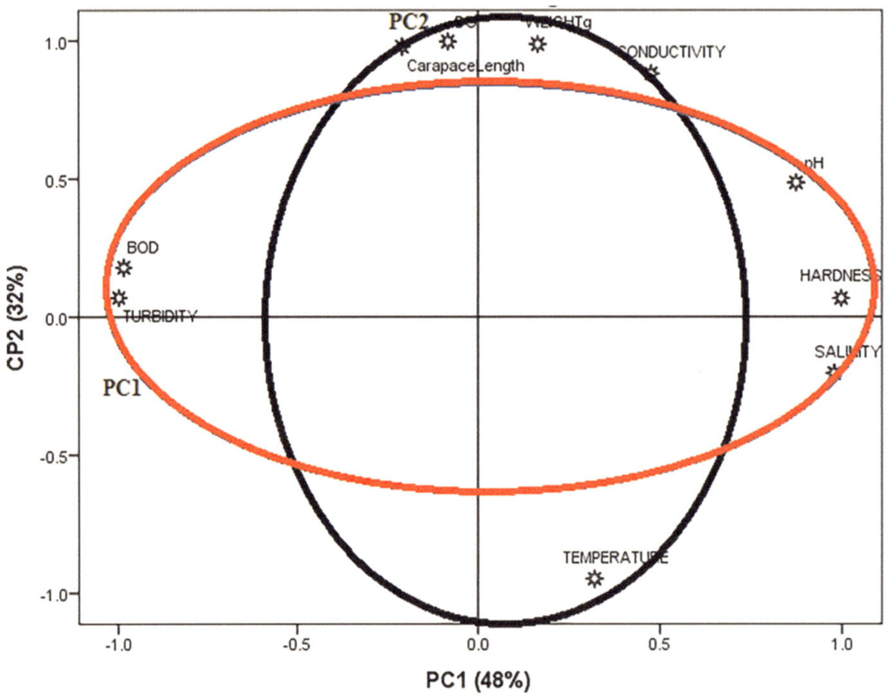

Figure 7.
Relationships (PCA) of the morphology of Callinectes marginatus *and water quality.*

boosting the oxygen level of the ecosystem for a period of time. Water temperature recorded in the study was within the optimal limit for general metabolism, growth performance and enhancement of aquatic species [22].

The turbidity, salinity and conductivity concentrations were within the permissible range for coastal marine waters. There were variations in the turbidity and conductivity concentrations across the stations which could limit the distribution and abundance of organisms that requires stable concentration for reproduction, growth and survival. The hardness, biochemical oxygen demand (BOD) and electrical conductivity (EC) of the water body were extremely higher than the recommended limit for coastal marine waters [19], hence the 'heavy pollution' status of the BOD was not surprising. According to Yan et al. [15] these high concentrations could be adduced to the availability of a large amount of organic materials, a large quantity of urban runoffs, aggregation of both solid and domestic wastes and high concentration of dissolved ions in the ecosystem. These may also be responsible for the 'moderate pollution' status of the study area.

Moreover, the single factor pollution index revealed 'slight pollution' status for the study area in terms of all the parameters except for BOD that indicated 'heavy pollution' and in turn influenced the comprehensive pollution index to indicate a moderately polluted aquatic ecosystem. This shows that the pollution of the aquatic ecosystem spans from the physically observed human-mediated activities (such as domestic discharges, industrial effluents etc.) in the environment. The pollution level of the ecosystem reveals the environment to be relatively poor for the sampled species and other aquatic organisms in the water body. This was also in consonance with the findings of Bolarinwa et al. [21] on the same coastal waters.

The population structure of decapod crustaceans in the coastal marine waters follows the order *N. hastatus>F. notialis > H. atlantica > M. macrobrachion > S. validus > O. africana > C. marginatus*. It revealed that the decapods were highly predominant with *Nematopalaemon hastatus* which represented about three-quarter of the sampled crustaceans. This can be attributed to the breeding pattern of the shrimps as it breeds all year round with peaks in June and November [23, 24]. Also, the shrimp's ability to tolerate and adapt to diverse ecological niche and environmental conditions could be responsible for its predominance. *Farfantepenaeus notialis* and *Holthuispenaeopsis atlantica* were the second and third most dominant species in the ecosystem corroborating the assertions of Powell [24] and Olawusi-Peters and Ajibare [10] that the two species were of secondary importance to *N. hastatus* in Nigerian coastal waters. The low population structure of the other sampled organisms could be attributed to the moderately polluted water quality of the ecosystem as revealed by the results of the principal components analyses. Factors such as pH, salinity, temperature, dissolved oxygen (which were slightly polluted) and BOD (which was heavily polluted) have shown to influence the distribution and abundance of organisms. For example, Deekae et al. [20] observed a positive relationship between the population of shrimps and temperature, salinity, dissolved oxygen etc. in Luubara Creek, Nigeria and strongly recommended the effective control of all activities in the ecosystem.

The morphological features of the sampled organisms were within the morphological ranges recommended by FAO [13] and Powell [24]. *Holthuispenaeopsis atlantica* recorded the highest morphological features when compared with the other species. The species had been conspicuously noticed and identified to be carnivorous when compared to other species, especially the penaeid shrimps. Powell, [24] further explained that the species feed on small crustaceans in the river mouth, thus accounting for its large size. Moreover, the total and body lengths were within the expected maximum size of 9–12 cm and 6–9 cm respectively for the species, thereby indicating the species to be in their adult stage. The most abundant species in the ecosystem (*N. hastatus*) showed diverse ranges of morphological features.

The average total length observed for *N. hastatus* in this study was within the range (4.87–9.71 cm) earlier stated by Ajibare et al. [25, 26] for white shrimps in the brackish waters of Ondo State, Nigeria. The shrimp's ability to tolerate wide ecological niche and environmental conditions could have been the reason for this [23]. Similarly, the observed morphological characteristics of *M. macrobrachion* in this study agree with the findings of Jimoh et al. [27] who studied female *M. macrobrachion* in Badagry creek, Nigeria and observed mean weight of 5.65 g. However, it was lower than the observations of Oyekanmi [28] and Ajibare et al. [29] who reported a body weight of 66.14 g and 76.25 g for *M. macrobrachion* in Asejire reservoir respectively. Alphonse et al. [30] also reported 15.74 g as the average weight of *M. macrobrachion* in Mono-River coastal lagoon system in the Republic of Benin. These variations may be as a result of the differences in the sex, season, location and pollution status of the habitats [25, 26]. Also, Daniels et al. [31] opined that widely distributed fish species have high variation in morphology features. This is substantiated by the higher carapace length recorded in the study when compared to the findings of Enin et al. [32] who worked on the population dynamics of estuarine prawns off the southeast coast of Nigeria and obtained CL of (1.67–2.01 cm).

The results of principal components analysis showed that there was a strong correlation within the morphological parameters of each studied species of shrimps and crabs. Also, Temperature, dissolved oxygen and the biological oxygen demand (which are important pollution indicators) had positive correlation and loaded in the same principal component with morphological parameters of each examined species. This indicated that the growth, size, morphology and abundance of each species of the shrimps (*F. notialis*, *M. macrobrachion*, *N. hastatus* and *H. atlantica*) and crabs (*C. marginatus*, *O. africana* and *S. validus*) might probably be affected by the temperature and the concentrations of DO and BOD since temperature reduces the DO (available to the biota), which in turn increases the BOD of the water [7, 22]. Also, pH, salinity, conductivity and turbidity (which had negative correlation with temperature, DO and BOD) contributed significantly to the survival, growth and abundance of each species of shrimps and crabs in the water body.

5. Conclusion

This study has used principal component analysis to examine the relationship between the tropical crustaceans and the environment and establish a baseline data on the eco-morphology of the coastal marine waters of Ondo state, Nigeria. The study revealed that the sizes of the decapod crustaceans were within the morphological range recommended by FAO and that all water quality parameters indicated slight pollution except BOD that indicated heavy pollution of the study area. However, the water quality (which was moderately polluted) can still sustain the biodiversity if the anthropogenic influence is regulated. The results of the principal components analysis (PCA) established that temperature, DO, and BOD had strong and positive correlation with the morphological parameters and therefore influenced the morphology/size of the crustaceans. Thus, improved management of the ecosystem is recommended in order to achieve healthy growth and survival of the aquatic species.

Funding disclosure

The authors of this research publication received no research funds/compensation from any organization. The research project and publication were sponsored by all the authors.

Competing interest statement

The authors have declared that no competing interest exists in the manuscript.

Author details

Adefemi O. Ajibare[1], Olaronke O. Olawusi-Peters[2] and Joshua O. Akinola[2*]

1 Department of Fisheries and Aquaculture Technology, Olusegun Agagu University of Science and Technology, Okitipupa, Nigeria

2 Department of Fisheries and Aquaculture Technology, Federal University of Technology Akure, Nigeria

*Address all correspondence to: joesheathakinola@gmail.com

References

[1] New MB. Farming Freshwater Prawns: A Manual for the Culture of the Giant River Prawn (*Macrobrachium rosenbergii*). Rome, Italy: FAO; 2002

[2] Sampaio ALA, Pagotto JPA, Goulart E. Relationships between morphology, diet and spatial distribution: Testing the effects of intra and interspecific morphological variations on the patterns of resource use in two Neotropical Cichlids. Neotropical Ichthyology. 2013;**11**:351-360

[3] Teixeira I, Bennemann ST. Ecomorfologia refletindo a dieta dos peixes em um reservató rio no sul do Brasil. Biota Neotropica. 2007;**7**:67-76

[4] Cunico AM, Agostinho AA. Morphological patterns of fish and their relationships with reservoirs hydrodynamics. Brazilian Archives of Biology and Technology. 2006;**49**(1): 125-134

[5] World Fish Center. The threat to fisheries and aquaculture from climate change. Penang, Malaysia: World Fish Center; 2007. p. 8. Available from: https://digitalarchive.worldfishcenter.org/bitstream/handle/20.500.12348/1594/ClimateChange2.pdf?sequence1=

[6] Morgan KM, Perry CT, Smithers SG, Johnson JA, Daniell JJ. Evidence of extensive reef development and high coral cover in nearshore environments: Implications for understanding coral adaptation in turbid settings. Science Reports. 2016;**6**:29616

[7] Ajibare AO, Ayeku PO, Akinola JO, Adewale AH. Plankton composition in relation to water quality of the coastal waters of Ondo state, Nigeria. Asian Journal of Fisheries and Aquatic Research. 2019;**5**(2):1-9

[8] Akinola JO, Olawusi-Peters OO, Akpambang VOE. Ecological hazards of Total Petroleum Hydrocarbon in Brackish water white shrimp *Nematopalaemon hastatus* (AURIVILLUS 1898). Egyptian Journal of Aquatic Research. 2019;**45**(3):205-210

[9] Akinola JO, Olawusi-Peters OO, Akpambang VOE. Human Health Risk Assessment of TPHs in Brackish Water Prawn (*Nematopalaemon hastatus*, AURIVILLUS 1898). Heliyon. 2020;**6**(1): 1-6

[10] Olawusi-Peters OO, Ajibare AO. Species richness, diversity and abundance of some Decapod Crustaceans in coastal waters of Ondo State, South West, Nigeria. International Journal of Fauna and Biological Studies. 2014;**1**(5):44-51

[11] Olawusi-Peters OO, Ajibare AO, Bello-Olusoji OA. Length-weight relationship and condition factor of shrimps in coastal waters of Ondo state, South West, Nigeria. International Journal of Fisheries and Aquatic Studies. 2014;**1**(3):137-142

[12] Olawusi-Peters OO, Akinola JO, Jelili OA. Assessment of heavy metal pollution in water, shrimps and sediments of some selected water bodies in Ondo State. Journal of Research in Agricultural Science. 2017; **5**(2):55-66

[13] FAO. Species Identification Sheets. Food and Agricultural Organization of the United Nations, (Volume VI). Department of Fisheries and Oceans, Canada; 1981. pp. 7–35

[14] American Public Health Association (APHA). Standard Methods for the

Examination of Water and Waste Water. 19th ed. Washington, DC; 1995

[15] Yan CA, Zhang W, Zhang Z, Liu Y, Deng C, Nie N. Assessment of water quality and identification of polluted risky regions based on field observations and GIS in the Honghe river watershed, China. PloS One. 2015;**10**(3):e0119130

[16] Li Q, Zhou Y, Li P. Evaluation of water quality status of coastal water in Sanya Bay in summer. Transactions on Oceanology and Limnology. 2010;**3**: 100-106

[17] Tao T, Yujia Z, Kai H. Water quality analysis and recommendations through comprehensive pollution Index Method-Take Qilu Lake as Example. Management Science and Engineering. 2011;**5**(2):95

[18] FEPA. Guidelines and Standards for Environmental Pollution Control in Nigeria. Federal Environmental Protection Agency; 2003. p. 238

[19] Fundamentals of Environmental Measurements (FEM). Conductivity, pH, Salinity, Total Dissolved Solids, Acidity and Alkalinity: Fondriest Environmental Inc. 2015. Available from: http://www.fondriest.com/environmentalmeasurements/parameters/water-quality/conductivity-salinity-tds

[20] Deekae SN, Abowei JFN, Chinda AC. Some physicochemical parameters of Luubara Creek, Ogoni Land. Niger Delta, Nigeria. Research Journal of Environmental and Earth Science. 2010;**2**(4):199-207

[21] Bolarinwa JB, Fasakin EA, Fagbenro AO. Physicochemical analysis of the coastal waters of Ondo State, Nigeria. International Journal of Research in Agriculture and Forestry. 2016;**3**(11):13-20

[22] Ajibare AO. Assessment of physico-chemical parameters of waters in Ilaje Local Government Area of Ondo State, Nigeria. International Journal of Fisheries and Aquatic Studies. 2014;**1**(5): 84-92

[23] Ofor CO, Kunzel T. Aspects of the fishery and reproduction of *Nematopalaemon hastatus* (Crustacea: Palaemonidae) in the Cross River Estuary, Nigeria. African Journal of Tropical Hydrobiology and Fisheries. 2013;**13**:1-4

[24] Powell CB. Fresh and Brackish Water Shrimps of Economic Importance in the Niger Delta. In: Proceedings of the 2nd Annual Conference of the Fisheries Society of Nigeria, Calabar, 24–27-January 1982. 1982

[25] Ajibare AO, Olawusi-Peters OO, Bello-Olusoji OA. Effects of seasons on the length-weight relationship and condition factor of two Brackish Water Shrimps in Ondo State, Nigeria. Nigerian Journal of Agriculture, Food and Environment. 2017;**13**(3):52-58

[26] Ajibare AO, Olawusi-Peters OO, Oyinlola KA. Microbiological load evaluation of white shrimp (*Nematopalaemon hastatus* Aurivillius, 1898) in the coastal waters of Ondo state, Nigeria. Agricultural Science and Technology. 2021;**13**(4): 425-431

[27] Jimoh AA, Clarke EO, Whenu OO, Anetekhai MA, Ndimele PE. Morphological characterization of populations of *Macrobrachium vollenhovenii* and *Macrobrachium macrobrachion* from Badagry Creek, Southwest Nigeria. Asian Journal of Biological Sciences. 2012;**5**:126-137

[28] Oyekanmi FB. Bio-Ecology of freshwater prawns Macrobrachium

vollenhovenii (Herklots, 1857) and Caridina africana (Kingsley, 1882) at Asejire Reservoir and Erin-Ijesa Waterfalls, Nigeria [doctoral thesis]. Abeokuta: University of Agriculture; 2011. p. 169

[29] Ajibare AO, Omobepade BP, Loto OO. Condition factor and length-weight relationship of berried African River Prawn (*Macrobrachium vollenhovenii)* in Asejire Reservoir, Nigeria. West African Journal of Fisheries and Aquatic Sciences. 2020; **1**(1):35-42

[30] Alphonse A, Youssouf A, Edmond S, Gildas-Gbaguidi MHA, Fiogbe ED. Meristic and morphological characterization of the freshwater prawn, *Macrobrachium macrobrachion* (Herklots, 1851) from the Mono River – Coastal Lagoon system, Southern Benin (West Africa): Implications for species conservation. International Journal of Biodiversity and Conservation. 2013;**5** (11):704-714

[31] Daniels SR, Stewart BA, Ridgway TM, Florence W. Carapace dentition patterns, morphometrics and allozyme differentiation amongst two toothed freshwater crab species (*Potamonautes warreni* and *P. unispinus*) (Decapoda: Brachyura: Potamonautidae) from river systems in South Africa. Journal of Zoology. 2001;**255**(3):389-404

[32] Enin U, Lowenberg U, Kunzel T. Population dynamics of the estuarine prawn (*Nematopalaemon hastatus* Aurivillius 1898) off the southeast coast of Nigeria. Fisheries Research. 1996;**26** (1–2):17-35

Chapter 3

Biological Characteristics of Some Fish Species in the Mekong Delta, Vietnam

Giang Van Tran and Quang Minh Dinh

Abstract

Vietnam, with a rich river system, many lagoons, and a long coastline, the fish fauna comprises both brackish and freshwater fish. The system of Mekong river is an extensive river located in the south of Vietnam. There are all kinds of brackish and freshwater fish in this river system. This chapter reveals data on the fish species composition and their roles in the Vietnamese Mekong delta. Moreover, the growth pattern and condition factor of some commercial fishes are also presented in this chapter. Their population structure and fishing status are also reported in this chapter.

Keywords: condition factor, goby, growth pattern, population structure, Vietnam

1. Introduction

Vietnam is one of the countries with a developed river system that comprises several extensive river systems, such as the Red river, the Ca river, the Dong Nai river, the Tien river, and the Hau river. Most of these rivers originate from other countries and flow into our country. According to statistics, about 2360 large and small rivers are widely distributed from the north to the south. With these characteristics, the fish system in Vietnam is very diverse and rich, with 1027 species classified into 427 genera, 98 families, and 22 orders. This fish system is considered diverse in species and high in biodiversity. The Mekong delta (VMD) is one of the two great deltas of Vietnam and is ranked 3rd globally [1]. The VMD is the last basin of the Mekong river with two main tributaries, the Tien and Hau rivers. This is an area with flat terrain and a dense system of rivers. On the other hand, VMD is also adjacent to the east sea and the Gulf of Thailand, with a coastline of nearly 700 km. Mekong river and rainfall mainly supply VMD's water source. The terrain in this area is favorable for the strong development of fishing and aquaculture from freshwater to brackish water [2]. The fish fauna in this area includes 332 species belonging to 77 families [3], including nearly 80 species of high economic value fish [4]. However, with the current exploitation of fish species, the number of some species is rapidly decreasing. Research on their biology is needed to supply appropriate data for the conservation and development of high-risk fish species. Therefore, the study "Biological characteristics of some fish species in the Mekong delta, Vietnam" provides these data.

2. Material and methods

This research was conducted at four sites covering a variety of aquatic environments. The first site has fresh water all year round in Cai Rang (Can Tho city, CRCT), the second site has brackish water due to saline intrusion in Long Phu (Soc Trang province, LPST); and year-round brackish water sites in Hoa Binh (Bac Lieu province, HBBL) and Dam Doi (Ca Mau province, DDCM). The biological characteristics of some investigated fish species include *G. aureus*, *G. giuris*, and *Glossogobius sparsipapillus*. The survey period lasted from January (2020) to March (2021). Fish samples were gathered using bottom nets with a net of $2a = 1.5$ cm. Collected fish were analyzed in the laboratory with the following steps: (1) classify founded on the outer morphological characterization [3], (2) sex discrimination based on genital spines [5], (3) measure morphological, and (4) anatomical parameters. Morphological parameters were estimated, including total length (*TL*, cm), and weight (*W*, g) (**Figure 1**).

Data on the *TL* and *W* of each fish sample was employed to specify the regression equation between length and weight of fish based on a formula: $W = a{*}TL^b$ (a and b were the coefficients). The coefficient b was employed to determine the growth pattern of the fish, for example, homologous growth when $b \neq 3$; unequal growth when $b \approx 3$, based on the research method of Froese [7]. The *CF* coefficient was determined by a formula, specifically, $CF=W/TL^b$ [8]. Where b was the growth coefficient (from the regression equation between the *TL* and *W* of fish).

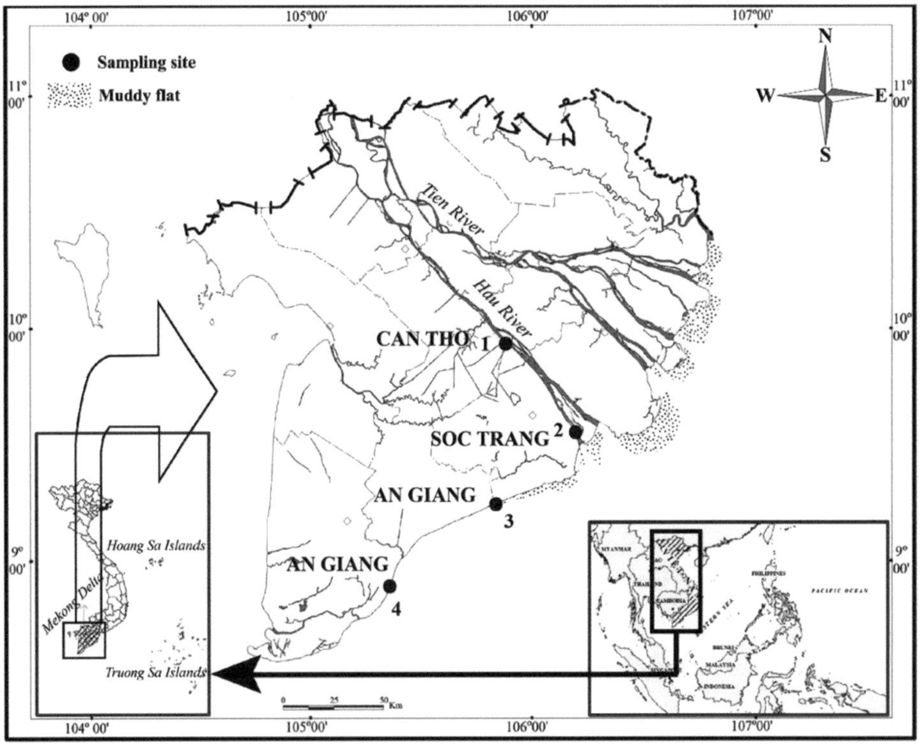

Figure 1.
Sampling location in the study area. 1: Cai RangCan Tho, 2: Long PhuSoc Trang, 3: Hoa BinhBac Lieu, 4: Dam DoiCa Mau (modified from Dinh [6]).

The collected length data were sorted into different length groups to determine the population parameters through the formula $L_t = L_\infty(1-e^{-K(t-t0)})$. Where L_∞ was the maximum asymptotic length that the fish could reach (cm), K was the growth rate of the fish, t was the age of the fish at time t; t_0 was the conjectural age at which the length of the fish was zero [9]. Length frequency data were included in the FiSAT II software to definite the L_∞ and K of the population using the ELEFAN I feature [10]. Length frequency data were normalized using the NORMSEP feature to serve as the basis for determining t_0 using the "Analysis of length-at-age" feature by [11].

The composite growth factor (Φ') was counted from the formula $\Phi' = \log K + 2*\log L_\infty$ [12]. The longevity was determined by the formula $t_{max} = 3/K$, where was the growth rate [13, 14]. The total dead factor (Z) was specified by a yield curve converted from length-frequency data [15]. Natural mortality (M) was determined by Pauly's [13] formula $\text{Log}M = -0.0066–0.279*\log L_\infty + 0.6543*\log K + 0.463/\log T$, where L_∞ was the maximum asymptotic length that the fish can reach (cm), K was the growth rate of fish, and T was the average annual surface water temperature. Mining mortality (F) and extraction coefficient (E) were determined according to the formula $F = Z–M$ and $E = F/Z$ [16].

The first catch length (L_c) was the length at which 50% of fish were caught and was determined by the yield curve conversion equation [9]. The Beverton & Holt [17] model was used to analyze the yield-to-addition (Y'/R) and biomass-to-addition (B'/R) models as the basis for determining the maximum yield (E_{max}), the optimal mining factor ($E_{0.1}$), and the mining factor at which B'/R was reduced by 50% ($E_{0.5}$). In addition, the ratio between L_c and L_∞ (isopleths) of fish was also analyzed. This data (L_c/L_∞) and the catch coefficient (E) were combined to determine the fishing status of the fish population established on the research method of Pauly & Soriano [18].

A t-test determined the variation of b and CF by sex and season. The changes of b and CF by length group and sampling site were employed in a one-way ANOVA. The population parameters were analyzed with FiSAT II.

3. Overview of the river system in Vietnam

Vietnam has a dense system of rivers with a total length of more than 41,900 km. Due to the rainy conditions, several rivers and streams have formed up to 2360 rivers and large and small canals [19]. In which, there are many river systems with large basins, such as Red, Ky Cung, Thai Binh, Ma, Bang Giang, Vu Gia—Thu Bon, Ca, Ba, Dong Nai, and Mekong rivers [20]. In 2011, the Mekong and Thai river systems covered an area of more than 1,167,000 km^2, however, 72% of the surface discharge in these basins originated outside of Vietnam [19]. The topography is steep in Vietnam along the northwest and southeast axis, causing surface water to concentrate mainly in the east, where all the major river basins are located. Meanwhile, the western mountainous areas are much drier, mostly with streams and small rivers [21]. Almost all major rivers in Vietnam originate from outside. The vast majority of rivers here usually flow in the direction of northwest-southeast and empty into the sea. However, the exception is Ky Cung and Bang Giang rivers. These two rivers flow in the southeast-northwest direction. All rivers originate in high mountains, so upstream rivers are often very steep. Therefore, in the rainy season, the river water flows strongly; when returning to the river delta, it becomes winding and meandering. With the main water source of rivers originating from outside the territory, it is difficult to control the amount of water [22].

4. Overview of the fish in Vietnam

With a rich system of rivers and lagoons, the fish fauna of Vietnam is very diverse. According to Nguyen [23], Vietnam has 1027 species, classified into 427 genera, 98 families, and 22 orders. Vietnam is one of 16 countries that are assessed as having high biodiversity, species diversity, and fish species. Until 2016, 290 new fish species were announced in Vietnam. Today, with the strong development of molecular markers and morphological studies, the composition of fish species in Vietnam is increasingly being fully and accurately determined, including the Mekong river.

5. Fishes in the Mekong Delta

Fisheries managers and scientists have updated info on fish diversity within the Mekong river delta [24, 25]. The dominance of marine characterizes fish composition in the western estuaries originated species with Engraulidae and fish family being dominant. In contrast, estuaries resident species (Pangasidae, Ariidae, Cyprinidae, Cynoglossidae, and Engraulidae) primarily contribute to fish composition in the Japanese estuaries. Spatio-temporal variations of species composition might be because of the hydrological regime powerfully influenced by the Mekong flows. There are 14, including inland protected areas (IPAs), with sizes varying from 500 hour angles to 14,605 ha. All of them are Melaleuca swamp forests or fresh marshes [26, 27]. The aquatic setting of those IPAs is variable and consistent with seasons, for example, nearly IPA's area is inundated in flooding season; however, canal systems within the IPAs contain water during the dry season. Therefore, environmental conditions amendment dramatically between two seasons. A recent study instructed that the dyke systems close to the IPAs cause poor quality of water in each season and, thus powerfully have an effect on aquatic life [28], which explains that solely black fishes, that is, genus *Anabas testudineus*, *Channa striata*, etc., survive such conditions, and that they are resident species. Within the flooding season, the inundated IPAs provide feeding and nursing ground for white and gray fishes (short and long migration fishes, such as Cyprinidae, Botiidae, Cobitidae, Siluridae, and Pangasiidae). Improving the dyke systems, that is, lowering the peak of the dyke system, enabling water exchangeably, enhancing water quality, and supplying migration routes for fishes. However, the inland protected areas (IPAs) presently target the protection of Melaleuca forest, grassland, and water birds [27, 29] and, consequently, the dyke systems and hydrological management operate to stop the fire, and going diversity of aquatic animals and migration routes being ignored. Surprisingly, data on fish diversity in these IPAs are poor or not available. Recently, our team created the assessment of fish diversity in two IPAs (Lang fractional monetary unit and U Minh Thuong), and knowledge has proven that the variability of fish from outside the IPAs (better water quality and migration routes) is way more extensive than that from within the IPAs for each season. Therefore, higher management practices of the IPAs might reach each readying and preserving fish diversity.

One of the prominent functions of the mainstream is to supply deep pools that provide vital shelters and spawning grounds for many indigenous and endangered species, that is, *Bosemania microlepis*, *Hypsibarbus malcolmi*, *Pangasianodon gigas,* etc. [30, 31]. Halls et al. [32] reported that deep pools in the VMD provide the dry season refuges for about 200 fish species. There are 23 deep pools in the VMD located mainly in Dong Thap and An Giang provinces and believed to be the refuges for many white fishes [33].

6. Fish growth, condition factor

In *Glossogobius aureus*, study results on the length and weight of 742 individuals in four study sites found a close relationship ($r^2 > 0.85$, in all cases). It was found that this was a fish with equal growth because, in most months, the growth coefficient *b* was equal to 3. However, in February, April, May, and June, there was a dominant growth pattern of a length overweight and 2 months November and December, which dominates mass over the span. This coefficient reached the highest value in Hoa Binh—Bac Lieu, while it was the lowest in Cai Rang—Can Tho. The environment and the food source in HBBL may be more suitable for fish growth than in the rest of the regions. This coefficient in the wet season is higher than in the dry season. This suggests that fish could have a higher growth rate in the wet season than in the dry season because of more abundant food sources. There were also statistically significant differences between mature and immature fish. Specifically, this value of the immature fish group was higher than the mature fish group. The growth coefficient *b* for males was 2.90 ± 0.04 SE, and for females was 2.82 ± 0.04 SE, and the whole population's coefficient was 2.85 ± 0.03 SE (**Figure 2**). This coefficient did not differ between males and females, showing that the growth of males and females was the same in the study sites. This coefficient for the whole population showed that this species belongs to the inequality growth group $b < 3$. The length of the fish tended to grow faster than the weight of the fish. This was similar to previous research on this species in the Con Tron area, Soc Trang province [35]. Also, there was a similar growth in *P. elongates* [36].

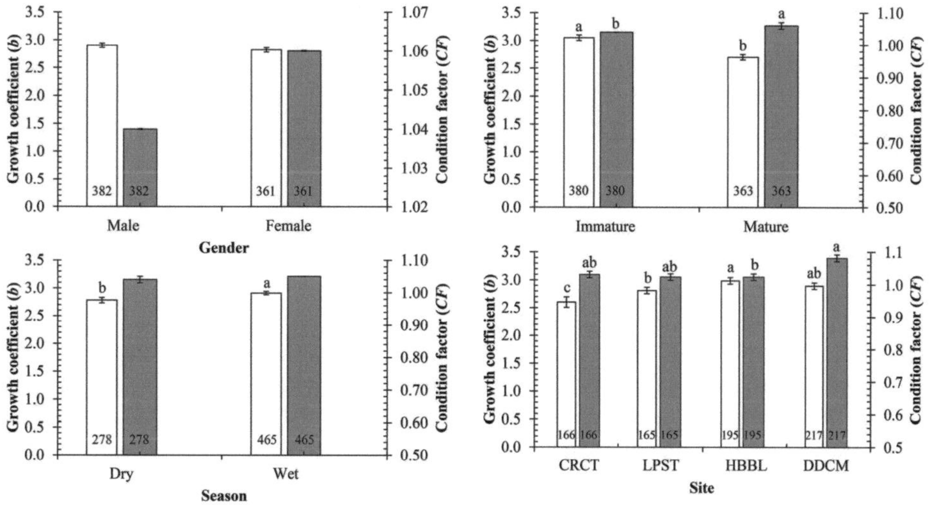

Figure 2.
Changed LWR and CF of sex, fish size, season, and sites in G. aureus. *(colorless columns: growth coefficient; gray columns: condition factor; number in each column: the number of individual fish; different letters (a, b and c) showed a significant difference; CRCT: Cai Rang—Can Tho; LPST: Long Phu—Soc Trang; HBBL: Hoa Binh—Bac Lieu; DDCM: Dam Doi—Ca Mau). Source: Phan et al. [34].*

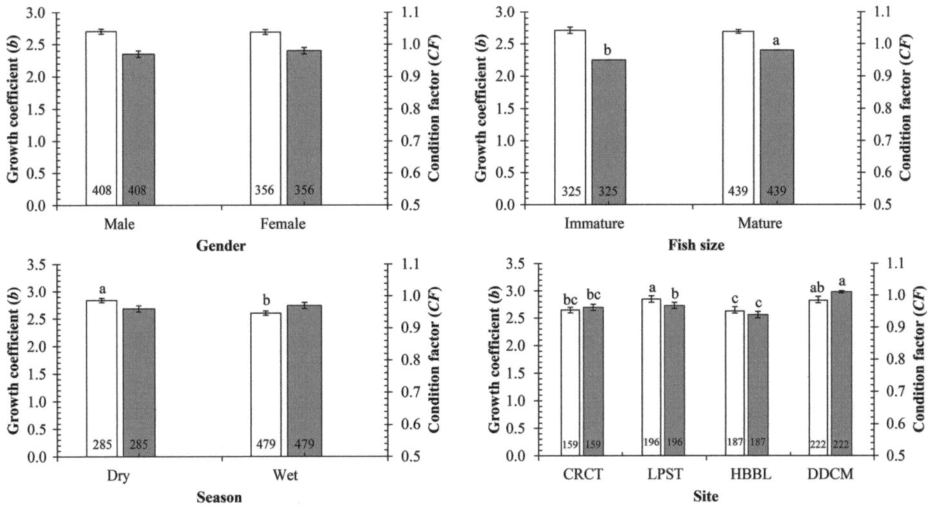

Figure 3.
Changed LWR and CF of sex, fish size, season, and sites in G. sparsipapillus (colorless columns: growth coefficient; gray columns: condition factor; number in each column: the number of individual fish; different letters (a, b and c) showed a significant difference; CRCT: Cai Rang—Can Tho; LPST: Long Phu—Soc Trang; HBBL: Hoa Binh—Bac Lieu; DDCM: Dam Doi—Ca Mau). Source: Truong et al. [37].

The study results on *G. sparsipapillus* showed a close relationship between fish length and weight in 764 individuals of this population at study sites ($r^2 > 0.7$, in all cases). Growth coefficient *b* in this species varied over the months. This was a species of an inequality growth group, and fish length grows faster than fish weight because, in most months, this coefficient was less than 3. The *b* value was highest in Long Phu—Soc Trang and lowest in HBBL. Similar to *Glossogobius aureus*, this species had *b* values difference that was not statistically significant ($p > 0.05$) in sex but was statistically significant between the dry season and the wet season (**Figure 3**). However, in terms of maturity, in *G. sparsipapillus*, there was no difference between mature and immature groups like fish below *G. aureus*. The growth coefficient *b* of the whole *G. sparsipapillus* population showed that this was a species of inequitable growth fish group, with the growing length faster than the fish weight with the mean value of 2.68 ± 0.03 SE. These results in two species were similar to many other fish species, such as *T. vagina* [38], *Boleophthalmus boddarti* [39], *Glossogobius giuris* [40], *Pseudapocryptes elongatus* [36], and the author's previous research on *G. aureus* [35] distributed in the Mekong delta.

Similar to *G. aureus* and *G. sparsipapillus*, *G. giuris* shows a positive relationship between fish weight and length ($p < 0.01$, $r^2 > 0.8$ for all cases). The growth coefficient of females was higher than that of males ($p < 0.05$), but this coefficient was not different between two groups of fish length ($p > 0.05$), as well as between wet and dry seasons ($p > 0.05$). Growth coefficients were similar across sex, fish size, and seasons ($b = 3, p > 0.05$). Unlike the two species above, the growth coefficient of this species (2.97 ± 0.04 SE) was equivalent to the equal growth value ($p > 0.05$) (**Figure 4**). This was similar to some other fish species, such as *B. boddarti* [39], *P. schlosseri* [42], *T. vagina* [38], *and Pd. elongatus* [36]. In the previous research of Phan et al. [41], *G. giuris* collected from Tra Vinh to Soc Trang and Bac Lieu also expressed the isometric growth pattern in most cases, except in October and

December 2016, the models of growth were negative allometry (2.79 ± 0.07 and 2.85 ± 0.07, respectively).

Condition factor (*CF*) in *G. aureus* differed over the study months (ANOVA, $p < 0.05$). There was no difference in sex and seasons (**Figure 2**). This showed that sex and environmental conditions in different seasons do not affect the fish's condition coefficient. However, there was a significant difference between the two groups of mature and immature fish. Specifically, this coefficient was higher in mature fish than in immature fish (**Figure 2**). At Dam Doi—Ca Mau, this index was the highest, followed by CRCT and LPST, the lowest was HBBL (**Figure 2**). Condition factor in *G. sparsipapillus* also changed over the study months ($p < 0.05$). *CF* was higher than in the other months in June, July, August, and September. Similar to *G. aureus*, the coefficient of this species was highest in DDCM and lowest in HBBL. There was no difference between males and females between dry and wet seasons ($p > 0.05$). But the mature fish group was higher than the immature group (**Figure 3**). The *CF* of *G. giuris* fluctuated monthly, reaching the highest value in November and the lowest in June (ANOVA, $p < 0.01$). The *CF* coefficient of females was higher than that of males, and that of the mature group was higher than the immature group (**Figure 4**). However, the *CF* coefficient of this fish in the dry season was equivalent to that of the wet season. Similarly, in the study of Phan et al. [34], *CF* values of this species were near the value of 1. Although the *CF* coefficients of the three species fluctuated by month, sex, and fish size, it was close to one threshold ($p > 0.05$) when considering the population as a whole. This showed that these fish were well adapted to the habitats.

Research results showed that these two species had similar growth patterns, with the growth pattern being negative allometry. With this growth pattern, the weight of these two fish species grew better than the fish length. The condition coefficients of *G. aureus* and *G. sparsipapillus* were equivalent to 1. This showed that these two fish species were well adapted to the environment.

7. Population structure

Length-derivative data for *G. aureus*, *G. giuris*, and *G. sparsipapillus* were converted to growth curves. Black and white bars in the von Bertalanffy growth curve showed a change in the groups' length and number over the study months. This equation of *G. aureus* had the parameters L_∞ = 30.44 cm, K = 0.51/year, $t_0 = -0.09$, and the von Bertalanffy growth curve equation for this population was L_t = 30.44 $(1-e^{-0.51(t\,+\,0.09)}L_t = 18.00(1 - e^{-0.53(t+0.10)})$. The von Bertalanffy growth equation of *G. giuris* parameters was L_∞ = 20.53 cm, K = 0.56/year, *and* $t_0 = -0.02$, and the von Bertalanffy growth curve equation, respectively, of the entire population $L_t = 20.53(1 - e^{-0.56(t+0.02)})$. *G. sparsipapillus* distributed in CRCT had the parameters of the equation were L_∞ = 16.53 cm, K = 0.78/year, and $t_0 = -0.10$. Meanwhile, these parameters at lower LPST were equal to L_∞ = 15.60 cm, K = 0.82/year, *and* $t_0 = -0.09$. The von Bertalanffy growth curve of the population in CRCT and LPST were $L_t = 16.53\,(1-e^{-0.78\,(t\,+\,0.10)})$ and $L_t = 15.60\,(1-e^{-0.82\,(t\,+\,0.09)})$, respectively (**Figure 5**).

From the yield curve converted from the length frequency, the total death coefficient (*Z*), the natural death coefficient (*M*), the logging death coefficient (*F*), and the extraction rate (*E*) of the *G. aureus* populations were 3.38/year, 1.16/year, 2.22/year and 0.66, respectively. Similarly, *Z*, *M*, *F*, and *E* of *G. giuris* populations were

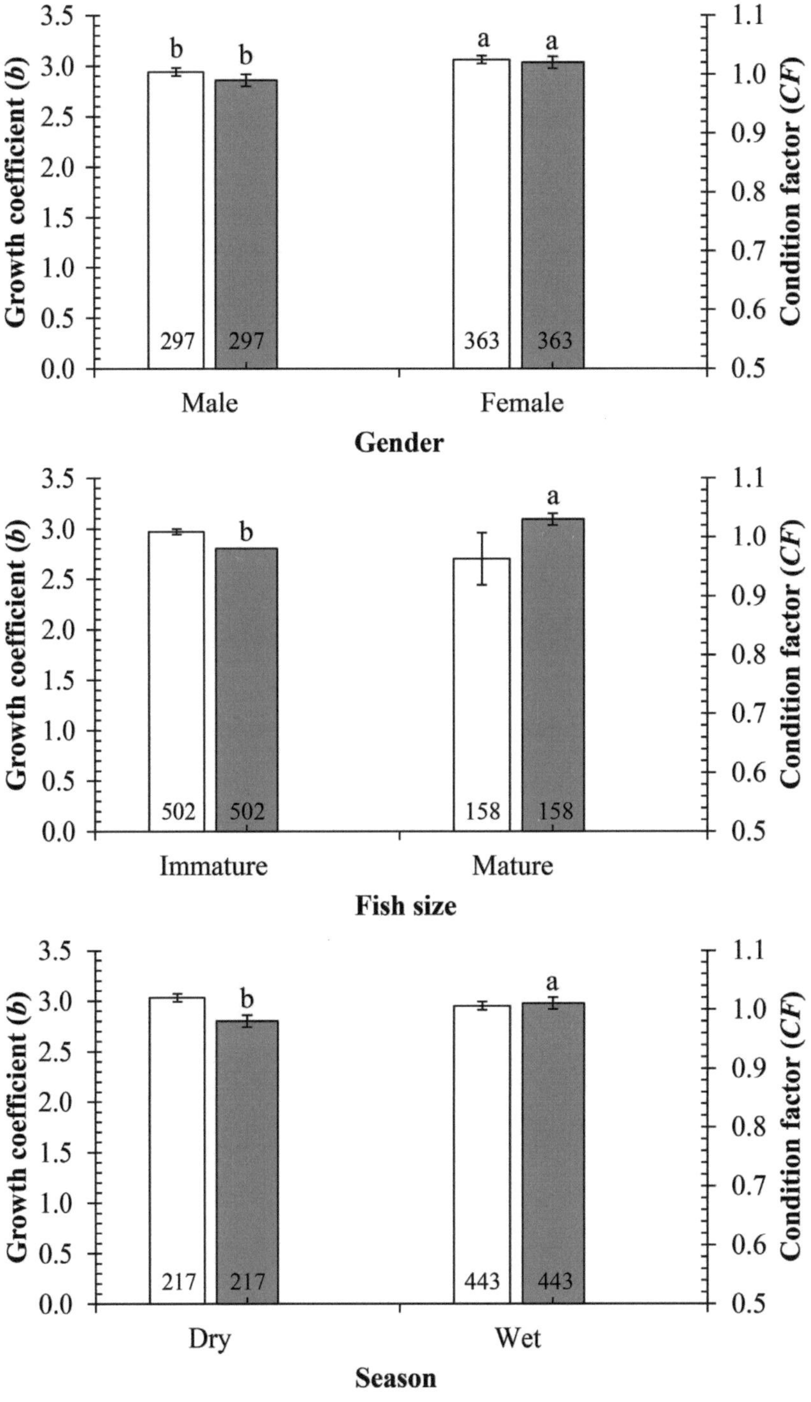

Figure 4.
Changed LWR and CF of sex, fish size, season, and sites in G. giuris. *(colorless columns: growth coefficient; gray columns: condition factor; number in each column: the number of individual fish; different letters (a and b) showed a significant difference). Source: Phan et al. [41].*

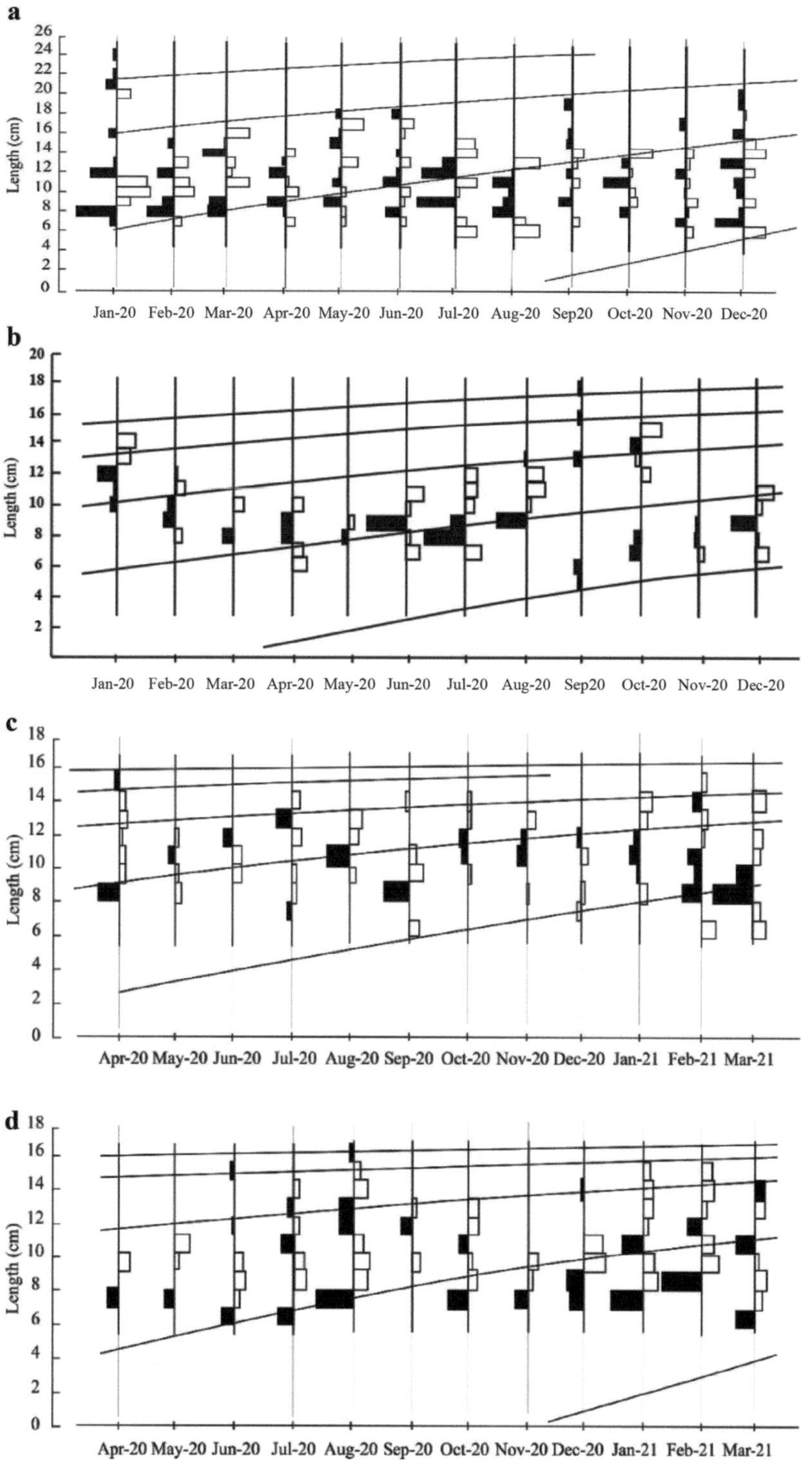

Figure 5.
The von Bertalanffy fish growth curve of G. aureus (a, n = 742, source: Dinh et al. [43]), G. giuris (b, n = 673, source: Dinh et al. [44]), and G. sparsipapillus (c: Cai Rang—Can Tho, n = 717, d: Long Phu—Soc Trang, n = 662, source: Nguyen et al. [45]) (the curves show the increase of fish length over time).

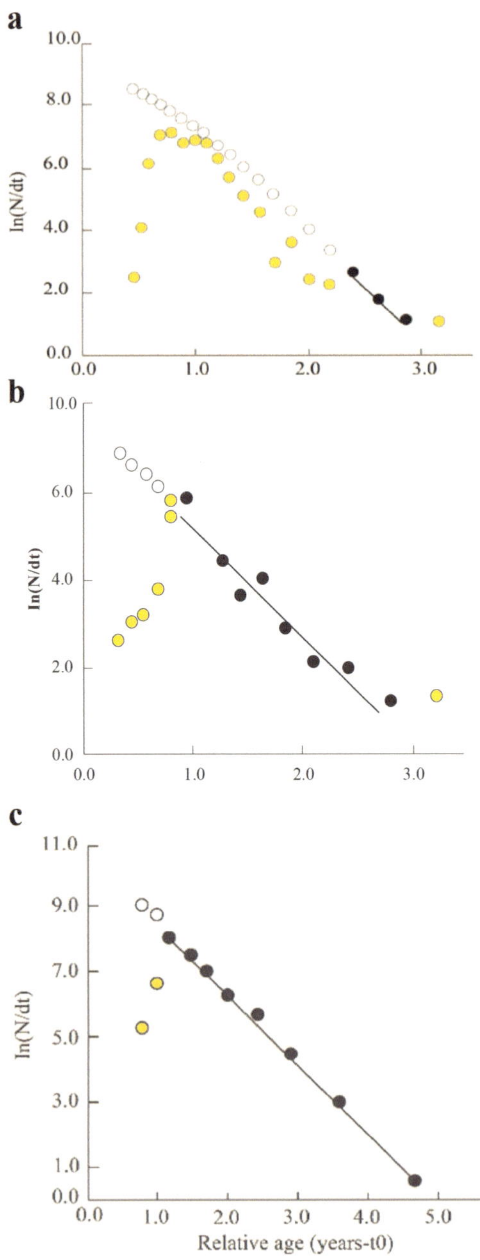

Figure 6.
The length converted catch curve G. aureus (a, source: Dinh et al. [43]), G. giuris (b, source: Dinh et al. [44]), and G. sparsipapillus (c: Cai Rang—Can Tho, d: Long Phu—Soc Trang, source: Nguyen et al. [45]).

3.17/year, 1.40/year, 1.77/year, and 0.56, respectively. In *G. sparsipapillus*, these coefficients were lower in CRCT (2.17/year, 1.49/year, 0.68/year, and 0.31/year) than in LPST (3.46/year, 1.68/year, 1.78/year, and 0.51/year). Similarly, the length at first capture in the CRCT population (5.07 cm) was shorter than in the LP population (8.01 cm) (**Figure 6**).

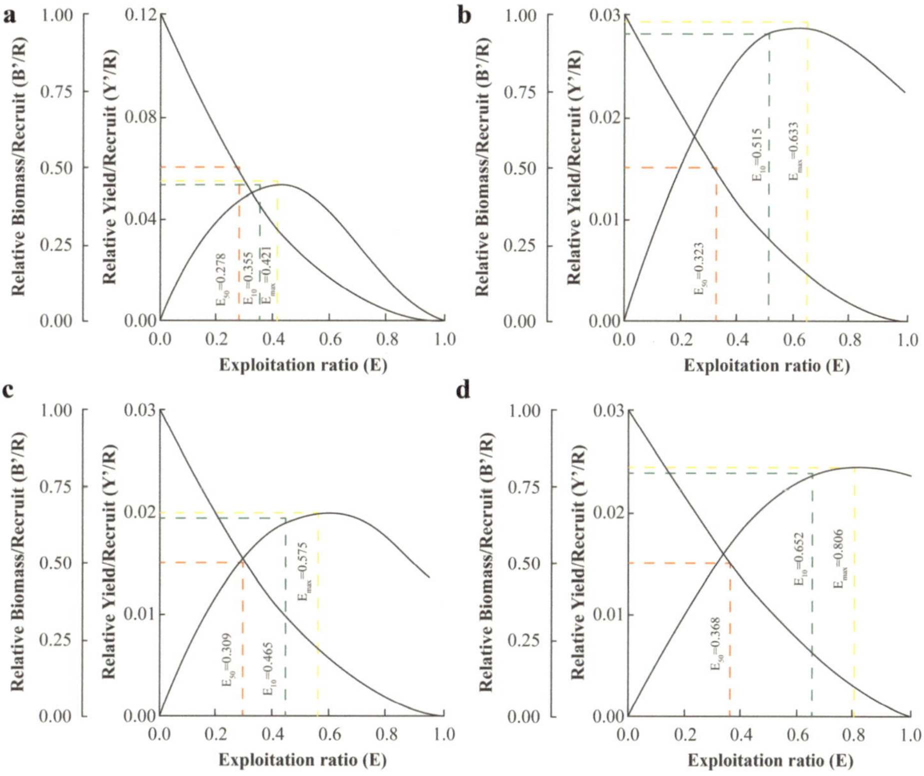

Figure 7.
The relative yield-per-recruit and relative biomass-per-recruit for G. aureus *(a, source: Dinh et al. [43]), G. giuris (b, source: Dinh et al. [44]), and G. sparsipapillus (c: Cai Rang—Can Tho, d: Long Phu—Soc Trang, source: Nguyen et al. [45]).*

After analyzing the results of analyzing the biomass and additional yield of two goby populations *G. aureus* at the four study sites, the maximum extraction coefficient (E_{max}) and the extraction rate were found optimal exploitation ($E_{0.1}$) and the exploitation coefficient that the population decreases by 50% ($E_{0.5}$) are 0.421, 0.355, and 0.278, respectively. Meanwhile, these coefficients in the *G. giuris* goby population were higher than those of the *G. aureus,* with these coefficients of 0.633, 0.515, and 0.323. These coefficients in *G. sparsipapillus* at the CRCT were 0.575; 0.465, 0.309, and at the LPST were 0.806, 0.652, and 0.368, respectively (**Figure** 7).

The growth coefficient (Φ') of the *G. aureus* population was estimated from the formula $\Phi' = \log K + 2\log L_\infty$ was 2.23 and the maximum lifespan of the population was 5.66 years. The rate of L_c/L_∞ in this species was 0.32. Similarly, in *G. giuris*, this rate was 0.36; its Φ' was 2.37 and t_{max} was 5.36. In *G. sparsipapillus,* the Φ' and t_{max} in the CRCT population were 2.19 and 4.81 yrs., respectively, and in the LPST population, they were 2.32 and 3.61 yrs., respectively. The L_c/L_∞ of this species was 0.43 in CRCT and 0.51 in LPST.

G. aureus, *G. sparsipapillus,* and *G. giuris*, had a higher growth rate than the other two species. However, all three species of goby had lower growth ratios than some other fish species in the same area, such as *Boleophthalmus boddarti* [46], *Parapocryptes serpersater* [47], *Stigmatogobius Pleurostigma* [48], *Pd. elongatus* Tran et al. [49], *Trypauchen vagina* [50]. Coefficient Φ' of *G. aureus*, *G. sparsipapillus,* and *G. giuris* was lower than other species, such as *Pd. elongatus* in the Mekong delta [49], *Glossogobius matanensis* in Indonesia [51] and *Pa. serperaster* in the Mekong delta [52]. It could be because the K and L_∞ of *G. giuris* were smaller than *Pd. elongatus* [49], *G. matanensis* [51] and *Pa. serperaster* [52].

Author details

Giang Van Tran[1] and Quang Minh Dinh[2]*

1 Department of Zoology, Hue University of Education, Hue University, Vietnam

2 Department of Biology, School of Education, Can Tho University, Vietnam

*Address all correspondence to: dmquang@ctu.edu.vn

References

[1] Coleman M, Huh OK. Major Deltas of the World: A Perspective from Space. Baton Rouge, LA: Coastal Studies Institute, Louisiana State University; 2012

[2] Le T, Nguyen MT, Nguyen VP, Nguyen DC, Pham XH, Nguyen TS, et al. Provinces and City in the Mekong Delta. In: Le T, editor. Geography of Provinces and Cities in Vietnam. Ha Noi: Education Publishing House; 2006. pp. 49-94

[3] Tran DD, Shibukawa K, Nguyen TP, Ha PH, Tran XL, Mai VH, et al. Fishes of Mekong Delta. Vietnam: Can Tho University Publisher; 2013. p. 174

[4] Thai NT. Research of the Biodiversity of the Mekong Delta Fish Family and their Change by Impacts of Climate Change and Social-Economic Development. Vietnam: Academy of Science and Technology; 2015

[5] Dinh QM, Le TTM. Reproductive traits of the duckbill sleeper *Butis butis* (Hamilton, 1822). Zoological Science. 2017;**24**:452-458

[6] Dinh QM. Aspects of reproductive biology of the red goby *Trypauchen vagina* (Gobiidae) from the Mekong Delta. Journal of Applied Ichthyology. 2018;**34**:103-110

[7] Froese R. Cube law, condition factor and weight-length relationships: History, meta-analysis and recommendations. Journal of Applied Ichthyology. 2006;**22**: 241-253

[8] Le Cren E. The length-weight relationship and seasonal cycle in gonad weight and condition in the perch (*Perca fluviatilis*). Journal of Animal Ecology. 1951;**20**:201-219

[9] Pauly D. Theory and Practice of Overfishing: A Southeast Asian Perspective. In: The international conference on the theory and application of length-based methods for stock assessment. Mazzara del Vallo: ICLARM. 1987. pp. 16-19

[10] Pauly D, David N. ELEFAN I, a BASIC program for the objective extraction of growth parameters from length-frequencies data. Meeresforschung. 1981;**28**:205-211

[11] Pauly D, Soriano-Bartz M, Moreau J, Jarre A. A new model accounting for seasonal cessation of growth in fishes. Australian Journal of Marine Freshwater Research. 1992;**43**:1151-1156

[12] Pauly D, Munro JL. Once more on the comparison of growth in fish and invertebrates. Fishbyte. 1984;**2**:1-21

[13] Pauly D. On the interrelationships between natural mortality, growth parameters, and mean environmental temperature in 175 fish stocks. Journal du Conseil. 1980;**39**:175-192

[14] Taylor CC. Cod growth and temperature. Journal du Conseil. 1958; **23**:366-370

[15] Pauly D, Moreau J, Abad N. Comparison of age-structured and length-converted catch curves of brown trout *Salmo trutta* in two French rivers. Fisheries Research. 1995;**22**:197-204

[16] Ricker WE. Computation and Interpretation of Biological Statistics of Fish Populations. Canada: Department of the Environment, Fisheries and Marine Service; 1975. p. 382

[17] Beverton RJH, Holt SJ. On the Dynamics of Exploited Fish Populations. London: Chapman & Hall; 1957. p. 533

[18] Pauly D, Soriano ML. Some practical extensions to Beverton and Holt's relative yield-per-recruit model. In: Maclean JL, Dizon LB, Hosillo LV, editors. The First Asian Fisheries Forum. Manila: Asian Fisheries Society; 1986. pp. 491-496

[19] Department of Water Resources Management. Overview of Vietna's Surface Water. Hanoi, Vietnam: Department of Water Resources; 2012

[20] Vietnam River Network. Map of 10 river basins in Vietnam. 2018

[21] Nguyen QV. Water Resource Fluctuations. Ministry of Natural Resources and Environment—Department of Water Resources Management; Hanoi, Vietnam. 2012

[22] Le BH. Declining Water Resources and the Risk of Water Insecurity in Vietnam. Vietnam: National Political Publishing House; 2013. p. 4

[23] Nguyen VH. Freshwater fish of Viet Nam. Vol. III. Ha Noi: Agriculture Publishing House; 2005. p. 655

[24] Hortle KG. Fishes of the Mekong—How many species are there. Catch and Culture. 2009;**15**:4-12

[25] Vidthayanon C. Field Guide to Fishes of the Mekong Delta. Vientiane: Mekong River Commission; 2008. p. 288

[26] Tran T. Transboundary Mekong River Delta (Cambodia and Vietnam). Vietnam: Springer Science; 2016

[27] Viet Nam Environment Protection Agency, (VEPA). Overview of wetlands status in Viet Nam following 15 years of Ramsar Convention Implementation. Hanoi, Vietnam 72. 2005

[28] Huynh VTM, Kurasaki M, Tran VT, Le KN, Avtar R, Rahman MM, et al. Effects of multi-dike protection systems on surface water quality in the Vietnamese Mekong Delta. Water Pollution and Fish Physiology. 2019;**11**:1-23

[29] Baran E, Guerin E, Saray S. Features of high fish productivity in the Mekong Basin: Case of 32 biodiversity hotspots. Report for the project "A climate resilient Mekong: Maintaining the flows that nourish life" led by the Nature Heritage Institute. WorldFish Center, Phnom Penh, Cambodia27. 2012

[30] Poulsen AF, Viravong OPS, Suntornratana U, Nguyen TT. Fish Migrations of the Lower Mekong River Basin: Implications for Development, Planning and Environmental Management. Phnom Penh: Mekong River Commission; 2002. p. 62

[31] Viravong S, Phounsavath S, Photitay C, Putrea S, Chan S, Kolding J, et al. Hydro-acoustic survey of deep pools in the Mekong River in southern Lao PDR and northern Cambodia. In: MRC Technical Paper. 2006;**11**:1-76

[32] Halls AS, Conlan I, Wisesjindawat W, Phouthavongs K, Viravong S, Chan S, et al. Atlas of deep pools in the Lower Mekong River and some of its tributaries. In: MRC technical paper. 2013. p. 31-69

[33] Vu VA, Nguyen ND, Nguyen MN, Hidas E. Deep pools as important fisheries habitats in the Mekong Delta. Fisheries ecology valuation and mitigation component. Mekong River Commission, Ho Chi Min City 46. 2007

[34] Phan GH, Dinh QM, Truong NT, Nguyen THD, Nguyen TTK. The intraspecific and spatio-temporal changes in growth pattern and condition factor of *Glossogobius aureus* inhabiting in

the Mekong Delta, Vietnam. Egyptian Journal of Aquatic Biology and Fisheries. 2021;**25**:591-599

[35] Dinh QM. Length-weight relationship of golden tank goby *Glossogobius aureus* Akihito & Meguro, 1975 in Con Tron River, Soc Trang Province. In: The 2nd National Scientific Conference on Marine Biology and Sustainable Development. Hai Phong, Vietnam: Natural Science and Technology Publisher; 2014. pp. 467-472

[36] Tran DD. Some Aspects of Biology and Population Dynamics of the Goby *Pseudapocryptes elongatus* (Cuvier, 1816) in the Mekong Delta. Malaysia: Universiti Malaysia Terengganu; 2008. p. 186

[37] Truong NT, Phan GH, Dinh QM, Nguyen THD, Nguyen TTK. Growth and condition factor of the commercial goby *Glossogobius sparsipapillus* living along Bassac River, Vietnam. AACL Bioflux. 2021;**14**:1695-1701

[38] Dinh QM. Growth pattern and body condition of *Trypauchen vagina* in the Mekong Delta, Vietnam. The Journal of Animal and Plant Sciences. 2016;**26**: 523-531

[39] Dinh QM. A preliminary study on length-weight relationship of the mudskipper *Boleophthalmus boddarti* in Soc Trang. Tap chi Sinh hoc. 2014;**36**: 88-92

[40] Dinh QM, Ly TV. Preliminary study result of length—Weight of tank goby, *Glossogobius giuris*, distributing in Soc Trang. Can Tho University Journal of Science. 2014; **2014**:220-225

[41] Phan GH, Le LTT, Dinh QM, Truong NT, Nguyen THD. Length-weight relationship, growth pattern and condition factor of *Glossogobius giuris* caught from coastal areas in the Mekong Delta. AACL Bioflux. 2021;**14**:1478-1485

[42] Dinh QM. Growth and body condition variation of the giant mudskipper *Periophthalmodon schlosseri* in dry and wet seasons. Tap chi Sinh hoc. 2016;**38**:352-358

[43] Dinh QM, Tran NQ, Tran DD. Some biological parameters of *Glossogobius aureus* population from the Mekong Delta. Iranian Journal of Fisheries Sciences. 2021;**20**:84-95

[44] Dinh QM, Phan YN, Tran DD. Population biology of the goby *Glossogobius giuris* (Hamilton 1822) caught in the Mekong Delta. Asian Fisheries Sciences. 2017;**30**:26-37

[45] Nguyen TTK, Dinh QM, Tran NS, Nguyen THD. Stock assessment of two populations of *Glossogobius sparsipapillus* (Osteichthyes, Gobiidae) in the Mekong Delta. The Egyptian Journal of Aquatic Research. 2021;**47**:401-407

[46] Dinh QM, Nguyen TTG, Nguyen TKT. Reproductive biology of the mudskipper *Boleophthalmus boddarti* in Soc Trang. Tap chi Sinh hoc. 2015;**37**: 362-369

[47] Dinh QM, Qin JG, Dittmann S, Tran DD. Reproductive biology of the burrow dwelling goby *Parapocryptes serperaster*. Ichthyological Research. 2016;**63**: 324-332

[48] Dinh QM, Nguyen NPD. Population and age structure of the goby *Stigmatogobius pleurostigma* (Perciformes: Gobiidae) from the Mekong Delta. International Journal of Aquatic Science. 2018;**9**:23-29

[49] Tran DD, Ambak MA, Hassan A, Nguyen TP. Population biology of the

goby *Pseudapocryptes elongatus* (Cuvier, 1816) in the coastal mud flat areas of the Mekong Delta,Vietnam. Asian Fisheries Sciences. 2007;**20**:165-179

[50] Dinh QM. Population dynamics of the goby *Trypauchen vagina* (Gobiidae) at downstream of Hau River, Vietnam. Pakistan Journal of Zoology. 2018;**50**: 105-110

[51] Mamangkey JJ, Nasution SH. Growth and mortality of endemic fish butini (*Glossogobius matanensis* Weber,1913) in Lake Towuti, South Sulawesi. Berita Biologi. 2014;**13**:31-38

[52] Dinh QM, Qin JG, Tran DD. Population and age structure of the goby *Parapocryptes serperaster* (Richardson, 1864; Gobiidae: Oxudercinae) in the Mekong Delta. Turkish Journal of Fisheries and Aquatic Sciences. 2015;**15**: 345-357

Chapter 4

Feeding Diversity of Finfish in Different Wild Habitat

Noor Us Saher, Raoof M. Niazi, Altaf Hussain Narejo, Noor Hawa, Abdul Hameed Baloch, Muhammed Tabish, Mussarat ul Ain, Faiqa Razi and Naureen Aziz Qureshi

Abstract

Sonmiani Bay has unique faunal diversity and distribution especially finfish as mangroves provides an imperative ecosystem which offer the shelter and protection to the associated organisms and care of their juveniles in bay limits. This study aimed to evaluate the diversity pattern according to physical and physiological responses and feeding habits (carnivorous and herbivorous) of finfish species in accordance with current habitat conditions in the Sonmiani Bay. A total of 4499, individuals of comprising 155 finfish species that represent 50 families were captured by these four (beach seines, purse seines, gill nets, and cast nets) nets during a twelve-month survey in a year. The distribution pattern of finfishes captured classified into four groups (tidal visitors, permanent residents, partial residents, and seasonal visitors) according to their patterns of distribution. Understanding of the true feeding behavior of organisms needs a more reliable and functional approach. The feeding ecology is not only functional for food and feeding behavior of fish as usually described by various tools and techniques of analytical research to take up more reliable details to explain the feeding biology in fish but also the indicator of habitat quality and status.

Keywords: food and feeding, finfish, lagoon water, wild habitat, feeding diversity

1. Introduction

Ocean cover up the 99.8% of earth's livable space but its resources are finite and fisheries have increased greatly in local aptitude, regional reach, and mechanical capacity throughout the world during the past half-century [1, 2]. Therefore, it has been observed that around 30% of marine fishery resources in the worlds are overexploited, 60% brutally utilized, and only 10% moderately exploited [3]. Many fish and shellfish species used the surface and intertidal regions of estuaries as nursing ground and transferred to the lagoon area with the help of tidal currents [4, 5].

The estuaries are the water mixing sites of marine and rivers, therefore, have a diverse number of compounds that are deposited here from a various ecosystems. To estimate the ecological parameters of an estuary with the help of decapods and finfish could be beneficial to assess the man-made impact on the ecosystem and their health risk on human health [6–8]. The estuaries can be divided into four categories as the natural estuaries, the lagoons, the tectonic estuaries, and the fjords [9–11]. Coastal lagoons, the permanent inland basins as connected to the nearby marine water (Sea or Ocean) by one or more inlets that remain open either continuously or periodically. The depth and size of lagoons usually depend on the adjacent sea level. The swampy coastal lagoons indicate the low sea level and when sea level is high, the water body appears as bays and coastal lakes. Only 13% of global coastline is comprised of lagoons [12] and thought to be distinct coastal area from estuaries [9].

The coastal lagoons are characterized as a sandy or muddy bottom, which are created and sustained by the deposition of sediments that carried by rivers, tides, currents, waves, and wind [13]. Mangrove finfish can be classified into four assemblages based on their distribution patterns: permanent residents, partial residents, tidal visitors, and seasonal visitors [14].

Mangrove canopy, an imperative feature of the Pakistan coastal areas and is most copious in the Indus Delta that constitutes about 97% of the total mangrove cover; whereas the 3% mangroves are found at three locations (Miani Hor, Jiwani at Gawatar, and Kalmat Hor) along the Balochistan coast. Mangrove habitats are highly productive and diversified areas as provide shelter to a number of invertebrates (crustaceans, polychaetes and mollusks) and also known as home of several commercial and non-commercial fishes [15]. The universal importance of this ecosystems is an essential habitat to maintain a variety of organisms and serve as feeding habitat, nurturing ground, and temporary and permanent residence to several finfish species and other invertebrate species [16–19]. Miani Hor is a protected mangrove environment and various authors explained Miani Hor as a highly variable place of commercial fauna as many fish, shrimp, and crabs are caught from mangroves which are carried out to the market and consumed by locals and the faunal diversity found in mangroves of Sonmiani [15, 20–22]. The studies on the significance of mangroves as feeding habitat for fish species and several commercially important macrobenthos have been provided by several authors worldwide [23–25].

Fishes are the important component of aquatic environment and have important contribution into the aquatic and terrestrial food chain. The analysis of fish feeding habits is essential for the functional role of those fishes which have not commercial importance and it's useful to understand the biological interaction, interspecies competition and to build trophic model by the diet composition [26, 27]. Food is a significant component of an organism for their survival and has major influence on the distribution, growth, reproduction migration rate, and behavior of an individual into the ecosystem [28, 29]. Food and feeding habits of an organism are important tool to understand the behavior of species, different aspect of energy flow, and relationship between predator and prey and consumer and widely address the trophic structure [30]. These can be analyzed by the morphological character of mouth shape, individual size, sex, age, locality of individual [31–33] and the composition and resources of the environment [34]. Availability of resources into an ecosystem often depends on the seasonal variation of climate, which lead to alter the nutrient levels and responsible to change the diversity and abundance of a community that influence the food habits of an individual [35].

The examination of food and determination of feeding habits for fish species is imperative to assess the place in the food web and biological role of any species in any ecosystems [36]. Information on the diet of any fish species can provide valuable guidance for the practices of the species culture and water body management, as needed for aquaculture and conservation purposes. Different fish species feed on wide range of food materials and obtain their nourishment from plants as well as animals. Depending upon the number of different type of food items consumed by them, fishes have been divided into two groups:

1. Stenophagic: Feeding on few type of food.

2. Euryphagic: Feeding on variety of food.

Food and feeding pattern of different fish is considered very important feature to help the selection of fish type for culture and farming. Fishes are herbivorous, carnivorous or omnivorous in nature, some fish groups are firmly herbivorous or carnivorous in habit, however various species remarkably adaptable in their food selection and feeding habits, therefore make use of the available food.

In aquatic ecosystem, among other nektons, fishes are a key consumer or top predator to occupy an obvious position in the trophic food web. The review of recent practice in feeding ecology of fish recognizes the need of directional efforts toward the assessment of descriptive ecology directly with the primary productivity in accordance with abundance of herbivore and carnivores finfish as primarily based on the diet information as mostly perceive directly through gut analysis or indirectly by computing some diet-based indices.

The widely used term feeding ecology explains the whole study about the feeding habit and acquired food of any particular species in relation with the habitat and tactics as animal adopts in specific environment to get its most enviable food through feeding or predation. In general, the tracing and occurrence of undigested food particles are recorded through stomach dissection [37, 38] for the qualitative and quantitative analyses. The examination of stomach contents along with various descriptive numerical techniques is used for estimation of diet and habits in aquatic animals [39]. There exists a different measure as described by various authors [37, 40–45] used as a handful tool to estimate stomach content and feeding habits of fish and likely used as an indicator of habitat status for any fish species.

An information of food components as ingested by the fish in natural habitat, quantitative analyses of gut content, the relevancy of morphological modification in the mouth or presence of any other supplementary body structure in relation with the food as intake by any particular fish are the most relevant areas to study the feeding ecology. However, the organic environment in the gut, sensational response for rejection and acceptance of food and responsive molecular signaling (**Figure 1**) are also important in feeding biology. The changes in position and structure of fish mouth are accountable for the food and feeding habits shown in **Figure 1**. Except mouth location and shape, further detailed study of various other factors like presence or absence of teeth, structure and number of teeth, mouth size, and presence of supplementary structure (spines, barbels), their location and modification can be helpful to determine the nature of food and habits of feeding in finfish species.

In previous studies, for the estimation of the fish stomach content different measures were adopted and used e.g. Index of Relative Importance [40]; Relation of total gut content weight with fish weight [41]; Feeding index [42]; Vacuity coefficient [44]; Visual assessment [40]; and Frequency of occurrences and volumetric

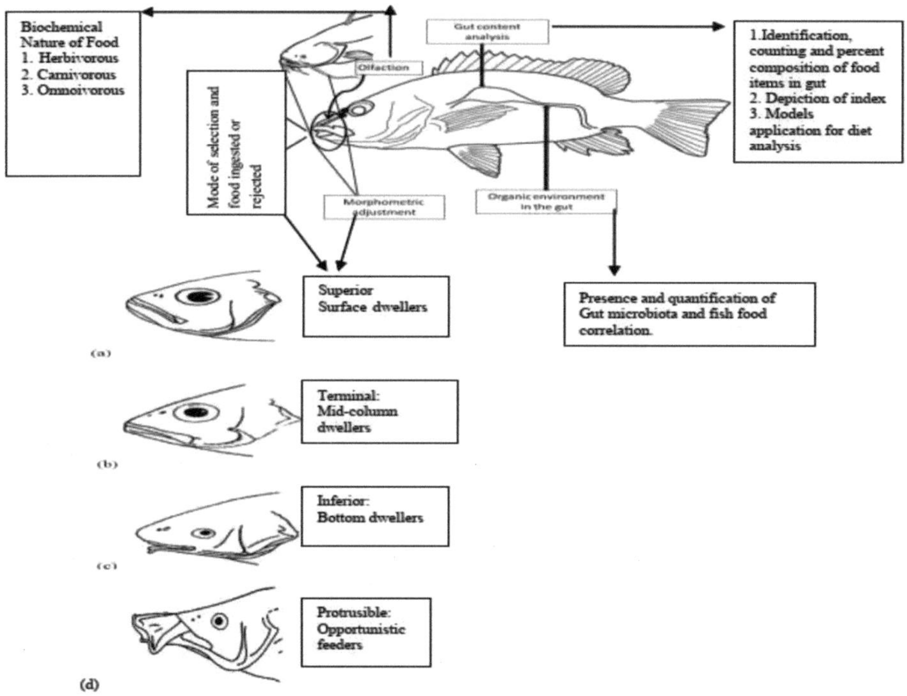

Figure 1.
A general description of areas needed to determine the feeding ecology in fish.

contribution [43]. There are also different measures for the analyses of stomach content except the mentioned methods. The different statistical and mathematical models are also present in the literature as usually applied for the description of food and feeding analyses; Electivity index [46]; Pianka's overlap index [47]; Hurlbert's diet breadth [48]; Levin's standardized index and Moritia's index [49]; Shannon index [50]; Repletion index [44]; Pelicice feeding activity index [51] and Saikia's diet breadth index [52].

The different finfish species like to live different type of habitats for their survivals and throughout their life stage including sand and mud substrates, oyster beds, water column, and sea grass. These species have wide range of feeding habits, and mostly depend on Crustaceans, Fishes, Bivalves, polychaetes as a diet which depend on the availability of prey, life stage locality, and species [53, 54]. There are enormous studies on the assessment of food and feeding habitat of different species of fishes [41, 55–58] in wild environment. The overview of feeding habits of mostly Finfish families is presented in **Table 1**.

In addition some particular and detailed information on the food and feeding habits of most commercial species are available [44, 111–127] in different region of the world show that there nature of consumers and feed on shrimp, Crab, other small fishes, polychaete and bivalve, that likely vary according to the size and age of the fishes [128, 129].

The present study determines the permanent and seasonal finfish diversity of lagoon waters and categorization of the fish feeding habit according to the available literature based on experimentation and food study in wild finfish species as collected throughout the year (each month) from lagoon waters of Sonmiani Bay.

Family	Distribution area	Feeding habit	References
Batrachoididae	Greater Caribbean (Florida to Belize), Mochima Bay.	Piscivorous	[59–61]
Gobiidae	Nigeria, Indo-West Pacific region, Persian gulf, Oman, Pakistan, India and Bay of Bengal, West Atlantic (Chesapeake Bay to Brazil and Georgia) Greater Caribbean (Florida to South Caribbean to Gulf of Mexico and Bermuda) Northern West Atlantic (Virginia to Suriname).	Juvenile-Herbivorous; Adult-Carnivorous	[59, 62–65]
	Baltic Basin	Carnivorous	[66]
	Norway	Carnivorous	[67]
	Pakistan	Cannibalistic and Carnivorous	[68]
	India	Cannibalistic	[69]
	Vietnam	Omnivore	[70]
	Slovakia	Carnivorous	[71]
Bothidae	Northern Sicilian coast, between Capo d'Orlando and Capo Calavà	Carnivorous	[72]
	Izmir bay	Carnivorous	[73]
Soleidae	Pakistan	Carnivorous	[74]
	Iran	Carnivorous	[75]
	Indo-West Pacific: Thailand to southern Japan, Borneo, Philippines, and Indonesia.	Invertivorous	[76]
Callionymidae	Indo-West Pacific: Persian Gulf to China, Indonesia, New Guinea, Borneo, Philippines and Southern part of southeast Asia, Western Central Pacific: Mediterranean Sea: Gulf of Genoa to the western Aegean Sea. Northwestern Australia and Papua New Guinea, Southern and eastern Black Sea.	Invertivorous	[77]; Fishbase.com
Cynoglossidae	Indo-West Pacific: Persian Gulf, Iran, Pakistan, India, Bangladesh, Myanmar, Sri Lanka, Philippines, Indonesia, Japan, and the north and east coasts of Australia.	Carnivorus	[75, 76]
Platycephalidae	Persian Gulf, India, Indo-West Pacific: Red Sea and East Africa to the Philippines, Eastern Mediterranean coast (Lebanon), North to Southern Japan and Korea, South to Northern Australia. Sicily, and strait of Messina.	Piscivorous and Invertivorous	[78–81]
Hemiscylliidae	Eastern Papua New Guinea; Arabian Sea to India, Pakistan, Sri Lanka, Singapore, Thailand, Vietnam, Indonesia, Taiwan, and Solomon Islands. Probably occurring in Korea and Japan.	Invertivorous	[82]
Synodontidae	Eastern Atlantic: Canary Islands, São Tomé Island, Morocco to Cape Verde, including the Mediterranean, and St. Helena Island, Western Atlantic: Florida, USA, Gulf of Mexico, Canada Indo-West Pacific: Northwest Atlantic: Somalia to Papua New Guinea, Japan and south to Indonesia.	Piscivorous	[83–87]
Dasyatidae	Indo-Pacific, India TO New Guinea, Japan, Atlantic and Caribbean, Northern South America, and West Africa.	Invertivorous	[76, 88, 89]
Lactariidae	Indo-West Pacific: East Africa to Southeast Asia, North to Japan, Queensland, Australia, Fiji.	Invertivorous	[90, 91]

Family	Distribution area	Feeding habit	References
Mugilidae	Cosmopolitan, Indo-Pacific: Pakistan, India, Japan, Red Sea to Samoa Eastern Pacific: California, USA to Chile, Japan to Australia, Western Indian Ocean: Western Pacific: South Africa, Nova Scotia, Canada to Brazil, Western Atlantic: Cape Cod to southern Gulf of Mexico; Eastern Atlantic: Bay of Biscay, Greater Caribbean Mediterranean Sea and Black Sea.	Planktivorous, Invertivorous and Detritivorous	[59, 79, 92]
Polynemidae	Indo-West Pacific: Pakistan, India, including Sri Lanka to Malay Peninsula, Persian Gulf to Papua New Guinea, Australia, West Atlantic (North Carolina to Brazil).	Piscivorous and Invertivorus	[59, 93]
Ariidae	Southeast Asia, Pakistan, India, West Indian Ocean (Polynesia and Japan), South New Guinea and Australia, Greater Caribbean and Gulf of Mexico	Omnivorous	[59, 79, 94]
Plotosidae	Indo-West Pacific: Red Sea, India, South Korea, East Africa, Samoa, Japan, Ogasawara Islands, Australia, Palau and Yap in Micronesia, Madagascar.	Invertivorous and Detritivorous	[79]
Clupeidae	Indo-West Pacific, Pakistan, India, Persian Gulf, Andaman Sea, Thailand, Indonesia, Vietnam, and Philippines Australia, the Caroline Islands, New Caledonia, Greater Caribbean, West Atlantic.	Planktivorous, Invertivorous	[59, 79, 95, 96]
Lutjanidae	Indo-West Pacific: Pakistan, India, Sri Lanka, Samoa, East Africa, and Australia. Ryukyu Islands, Solomon Islands, New Guinea, Indonesia, the Philippines, and China.	Invertivorous and Carnivores	[96]
Leiognathidae	Indo-West Pacific: India, Pakistan, New Guinea; Northern Japan; Southern Australia, Sri Lanka, Indonesia and Philippines, Red sea	Planktivorous, Invertivorous	[79, 96]
Sillaginidae	Western Indian Ocean: Arabian Sea, Pakistan, India, Indo-West Pacific, Eastern Africa and northward to Korea.	Invertivorous	[79, 91, 97]
Terapontidae	East Africa to Samoa, Red Sea, Japan, Australia, and Lord Howe Island.	Omnivorous and Invertivorous	[98]
Harpadontidae	Indo-West Pacific: Somalia, New Guinea, Japan, and Indonesia.	Carnivorous	[99]
Carangidae	South Africa, North to Red sea, Persian Gulf, India, Pakistan, Sri Lanka, Thailand, Japan, Indonesia to Fiji, New Zealand, Australia, Gulf of Papua.	Piscivorous and Invertivorous	[100]
Haemulidae	Indo-West Pacific: Pakistan to Sri Lanka, Western Pacific, China, Australia. Greater Caribbean (Gulf of Mexico-Guyana) and West Atlantic, Gulf of Suez Mozambique and South Africa.	Invertivorous (Nocturnal) and Carnivores	[59, 94, 96, 101]
Sparidae	South Africa, Mozambique, India, Red Sea, Persian Gulf, Malaysia (Penang Island and Langkawi Island, near Singapore). Northern Arabian Sea, Pakistan, Persian Gulf, Oman.	Carnivores, Invertivorous	[102, 103]
Sciaenidae	Indo-West Pacific, west to India and Sri Lanka, east to New Guinea, Pakistan, China and Australia, Greater Caribbean and Northern Western Atlantic (New York, Gulf of Mexico, Massachusetts, Argentina, Maine, Chesapeake Bay and Brazil).	Carnivores and Invertivorous	[59]

Family	Distribution area	Feeding habit	References
Serranidae	Portugal, Angola, including the Mediterranean, Pakistan, Indian Ocean: Northern India, Gulf of Aden to Sri Lanka, Persian Gulf, Red Sea. Greater Caribbean (North Carolina to Suriname and Gulf of Mexico).	Carnivores, Piscivorous, and Invertivorous	[59, 96]
Ophichthidae	Northern India, Pakistan, Gulf of Aden to Sri Lanka, Persian Gulf, Red Sea.	Carnivores, Piscivorous, and Invertivorous	[59, 79]
Terapontidae	India, Pakistan, Red Sea and East Africa, Japan, Taiwan, New Guinea, Indonesia, Malaysia, and Australia.	Carnivores	[104]
Gerreidae	Indo-Pacific: East Coast of India, Pakistan and Sri Lanka, Bangladesh, Myanmar, and Thailand. Bermuda and Florida, USA; Bahamas, northern Gulf of Mexico, South American coast to Rio de Janeiro, Brazil.	Invertivorous	[91]
Narcinidae	Pakistan, Western Indian Ocean, Gulf of Oman, Gulf of Aden.	Invertivorous	[101]
Mullidae	Indo-West Pacific: Red Sea to New Caledonia, Japan, Mozambique, Indo-Malayan region, Northern Australia, Fiji, east Mediterranean, Suez Canal.	Invertivorous	[105];
Scatophagidae	Indo-West Pacific; Fiji, Indonesia, Philippines.	Invertivorous and Detritivorous	[76]
Ambassidae	Pakistan; India, East Africa, South Africa, Western Pacific: Thailand, Indonesia and Philippines, New Guinea, Ryukyu Islands.	Planktivorous	[76, 101]
Gymnuridae	Indian Ocean and Central Pacific: India Pakistan Sri Lanka, Indonesia, Singapore and Thailand, West Atlantic: USA, Gulf of Mexico, extended to the Caribbean Sea, Central Pacific: Gulf of California and Panama.	Invertivorous	[106]
Belonidae	Pakistan, India, Persian Gulf, and Sri Lanka, Southern China, Northern Australia, and Philippines.	Piscivorous and Invertivorous	[107]
Chirocentridae	Pakistan	Carnivorous	[94]
Engraulidae	Gulf of Aden, Persian Gulf, Pakistan, India, Bay of Bengal Andaman Islands, Indonesia, Thailand, Taiwan Island, South to Northern Australia; Fiji and Tonga.	Planktivorous and Invertivorous	[79, 91, 94]
Hemiramphidae	India, Pakistan, Philippines, New Guinea, New Caledonia, and northern Australia; Peninsula of Japan, Indonesia, New Guinea Gulf of Thailand, and western Polynesia.	Omnivorous	[79, 91]
Sphyraenidae	India, Pakistan, Red Sea, East coast of Africa to Hawaii; Marquesas and Tuamotu islands, USA to Uruguay, New Jersey, Persian Gulf, Red Sea, East Africa; Japan to New Caledonia, Hawaii and Tahiti. California, USA to Ecuador and Peru.	Carnivorous	[96]
Lethrinidae	Red Sea, East Africa, Seychelles, Maldives, Sri Lanka, Andaman, Indonesia, Philippines, South Japan, Northeast Australia, New Guinea, Fiji, Vanuatu.	Invertivorous and Carnivorous	[108]
Nemipteridae	Indo-West Pacific: India and Pakistan, Bay of Bengal, southern Japan to Indonesia and northwestern Australia.	Planktivores and Invertivorous	[91, 94]

Family	Distribution area	Feeding habit	References
Trichiuridae	Cosmopolitan, Pakistan (Sindh and Balochistan), Indo-Pacific (Arabian Sea-North eastern, Indian ocean-northwest and central east) and Atlantic areas.	Carnivorous	[109]
Scombridae	Pakistan, Atlantic and Mediterranean, Tropical and Subtropical seas.	Carnivorous	[96]
Stromateidae	West Atlantic (Chesapeake Bay to Argentina).	Juvenile-Planktivorous Adult-Invertivorous	[59, 110]
Tetraodontidae	Red Sea, Pakistan, India, Persian Gulf to South Africa, Japan to Australia. South coast of South Africa, Indonesia, Papua New Guinea, Caribbean Greater.	Carnivorous	[59, 96]
Triacanthidae	Indo-West Pacific: Gulf of Oman, Arabian Sea, Persian Gulf, Bay of Bengal, and Japan China, South China Sea including Gulf of Thailand, Indonesia, Philippines, northern Australia.	Invertivorous	[76]

Table 1.
A detailed review of literature on the feeding habits of various finfish families.

2. Materials and methods

The lagoon area of Sonmiani Bay (Miani Hor) is located at the east coast of Balochistan and 90 km far from Karachi city [21, 130]. It is an estuary system having various islands, intertidal mudflats, and an extensive mangrove swamp. It extends up to 60 km and is widespread up to 7 km. The twisted and complicated water bodies

Figure 2.
The coastal area of Sonmiani bay showing the sampling area with location (stars) of the net activities performed (yellow stars = gill net, green stars = cast net, violet star = commercial beach seine, and red stars = purse seine).

connect it to the Arabian sea [131]. It is known as the largest bay in Pakistan. It spreads up to 363.3 square kilometers and about 80 km wide shelf area [132]. The rainwater runoff here from Porali and Windor rivers during the rainy season. This bay is extending up to 60 km in length and 7 km wide, twisted, and contorted body of water (**Figure 2**). Churna and Kiou islands are two islands present offshore and connected with intertidal mudflats, muddy beaches, mangrove forests, sandy beaches, and Hub River [131] characterizes Miani Hor. Three mangrove species have been reported from Pakistan (i.e. *R. mucronata; C. tagal* and *A. marina*). Miani Hor is the only region in Pakistan, where all these species are present together in a natural environment [133].

2.1 Field sampling

For the sampling of Finfish species, four sites were selected for a period from December 2001 to November 2002. Plate 1, shows the sampling area of this study which was comprised of the mouth of the bay, the front area of sand dune, mangrove creeks off Damb, and mangrove creeks of Bhaira village. The sampling site was muddy with some patches of sandy bottom. This helps to operate all types of nets in this area.

2.2 Fishing methods

The fishing methods/ nets adopted by local fishermen in *Miani Hor* (**Figure 3**), were also used for this study described as follows:

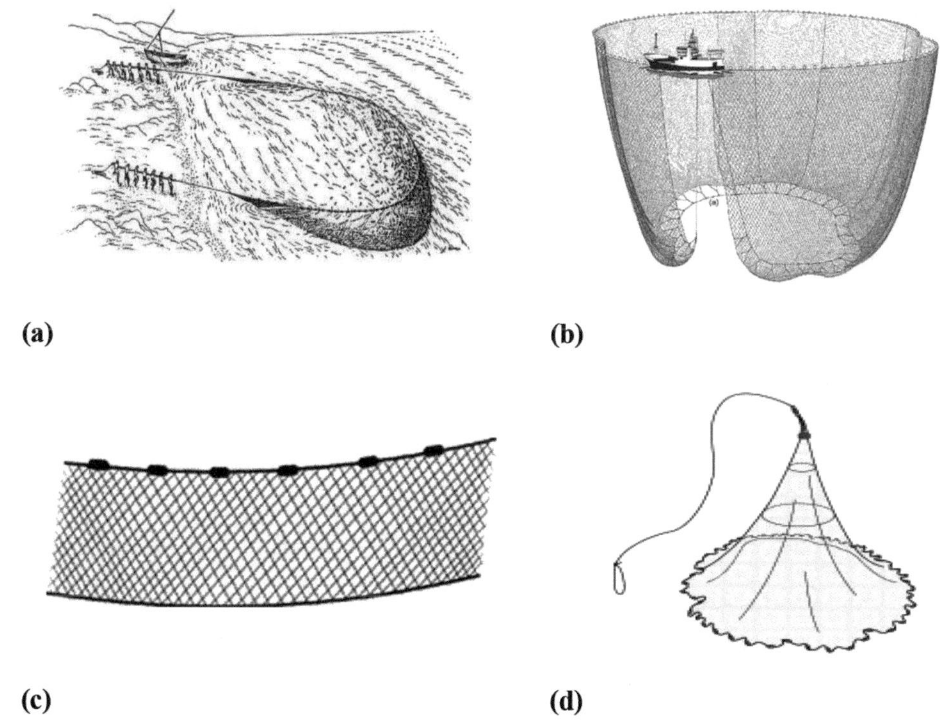

Figure 3.
The diagrammatic view showing four different types of gears used for study purposes (a) beach seine (b) purse seine (c) gill net (d) cast net.

2.2.1 Beach seine (Cada)

The beach seine is a well known commercial net used in Sonmiani Bay creeks. The net was used about 55.5 meters long and 9.2 meters wide at the center and at the corners it was about 7.4 meters wide curtains with 1 cm^2 mesh size. This methodology was used to operate by approximately 6–8 persons. This type of net is commonly used to catch small-sized pelagic and demersal fishes at high tide.

2.2.2 Purse seine (Katra)

The major commercial tool used to catch the sardine fisheries along the Sonmiani Bay (**Table 2**). It is about 250x100 meters curtain operating with two boats and with approximately 1 centimeter square mesh size. To capture a dense and mobile school of pelagic fish, the purse seine is effective.

2.2.3 Gill net (Tukri)

A type of commercial net provides major support for the shrimp fisheries of the Sonmiani bay (**Table 2**). Total size of Gill net used in the collection was 30.48 long and 3.65 meter wide with 1 square inch in mesh size. The gillnet is consisting of a web and a rectangular frame. It hangs into the water like a wall with the help of a head rope and foot rope.

2.2.4 Cast net (Goal-jal)

This is the type of non-commercial fisheries used by a fisherman in the coastal villages of the Sonmiani bay and support the domestic livelihood of the local people (**Table 2**). The cast net is a technical netting process and was used by only one person who was hired during collection time.

2.3 Laboratory analyses

2.3.1 Taxonomic identification

In the laboratory, the catch brought into the laboratory, washed, sorted according to the fish group and kept in marked polyethylene bags in deep freezer for subsequent study. The all collected finfish were identified up to the lowest taxonomic level. The fish species were identified with the help of field guide provided by Bianchi [91] and FAO Fish identification sheets by Psomadakis et al. [94].

No.	Sites	Nets	Status
1	Off Damb mangrove creeks	Beach seine, Gill net	Commercial
2	Bhaira village mangrove creeks	Gill net	Commercial
3	Mouth of the bay	Purse seine	Commercial
4	Damb Jetty Area	Cast net	Non-commercial
5	Sand dunes front	Cast net	Non-commercial

Table 2.
The status of different net types used to catch the marine fauna from the selected research sites of Sonmiani Bay.

3. Results

3.1 Finfishes diversity and habitat found in the Sonmiani Bay Area

During the current research total 155 finfish species from 49 families were collected and identified. Different type of fishing gears was used for the collecting of fishes "beach seine, purse seine, gill net, and cast net" in various habitat like shore line, pelagic and benthos pelagic area and mangrove area of Sonmiani Bay.

Because various finfish acquire varied behavior with habitat, all collected species were subjected to ecological factors such as habit and manner of feeding with reference to their habitat, morphological modifications and habitat adaptation. Mangrove finfish can be classified into four assemblages based on their distribution patterns: permanent residents, partial residents, tidal visitors, and seasonal visitors [14]. These variances are related to food and shelter. Bottom-dwelling fish with limited visibility live in the muddy mangroves. Following are four types of finfishes of the Sonmiani Bay area:

3.1.1 Permanent finfishes of Sonmiani Bay

Allenbatrachus grunniens of the Batrachoididae family and *Oligolepis acutipennis* of the Gobiidae family are examples of permanent finfish fauna (**Table 3**). Mudskippers are mostly found on the soft mud flat due to their typical adaptation such as pectoral

Permanent Finfishes of Sonmiani Bay		
Family	**Species**	**Common name**
Batrachoididae	*Allenbatrachus grunniens*	Grunting toadfish
Gobiidae	*Oligolepis acutipennis*	Sharptail goby
	Oplopomus oplopomus	Spinecheek goby
	Oxyurichthys microlepis	Maned goby
	Oxyurichthys papuensis	Frogface goby
	Trypauchen vagina	Burrowing goby
	Bolephthalmus sp.	Goby
Seasonal finfishes of Sonmiani Bay		
Family	Species	Common name
Bothidae	*Pseudorhombus elevatus*	Deep flounder
Soleidae	*Euryglossa orientalis*	Oriental sole
	Solea elongata	Elongate sole
Cynoglossidae	*Cynoglossus bilineatus*	Roundhead tongue sole
	Cynoglossus dispar	Carrot tongue sole
	Cynoglossus dubius	Carrot tongue sole
	Cynoglossus puncticeps	Speckled tongue sole
	Paraplagusia bilineata	Double lined tongue sole
	Cynoglossus sp.	Tongue sole
Platycephalidae	*Cociella crocodila*	Crocodile flathead
	Platycephalus indicus	Bartail flathead

Permanent Finfishes of Sonmiani Bay		
Family	**Species**	**Common name**
Callionymidae	*Callionymus marleyi*	Sand dragonet
Synodontidae	*Saurida undosquamis*	Brushtooth lizardfish
Dasyatidae	*Himantura gerrardi*	Stingray
Demersal waters dweller finfishes of Sonmiani Bay		
Family	Species	Common name
Mugillidae	*Liza carinata*	Keeled mullet
	Liza subviridis	Greenback mullet
	Liza tada	Tade gray mullet
	Valamugil cunnesius	Longarm mullet
	Valamugil seheli	Blue spot mullet
	Valamugil speigleri	Speigler's mullet
	Liza sp.	Liza Juvenile
Polynemidae	*Eleutheronema tetradactylum*	Fourfinger threadfin
	Polynemus heptadactylus	Seven thread tassel fish
	Polynemus indicus	Indian threadfin
Ariidae	*Arius arius*	Threadfin sea catfish
	Arius caelatus	Engraved catfish
	Arius crossocheilus	Roughback sea catfish
	Arius dussumieri	Blacktip sea catfish
	Arius maculatus	Spotted catfish
	Arius platystomus	Flatmouth sea catfish
	Arius sagor	Sagor sea catfish
	Arius sona	Sona sea catfish
	Arius subrostratus	Shovelnose sea catfish
	Arius tenuispinis	Thinspine catfish
	Arius thalassinus	Giant catfish
Plotosidae	*Plotosus lineatus*	Striped eel catfish
Clupeidae	*Anodontostoma chacunda*	Chacunda gizzard shad
	Nematolosa nasus	Bloch's gizzard shad
Lutjanidae	*Lutjanus russellii*	Russell's snapper
	Lutjanus lunulatus	Lunartail snapper
Leiognathidae	*Gazza minuta*	Toothpony
	Leiognathus berbis	Berber ponyfish
	Leiognathus blochi	Tow-blotch ponyfish
	Leiognathus daura	Goldstripe ponyfish
	Leiognathus equulus	Common ponyfish
	Leiognathus fasciatus	Striped ponyfish
	Leiognathus splendens	Splendid ponyfish
	Secutor insidiator	Pugnose ponyfish

Permanent Finfishes of Sonmiani Bay		
Family	**Species**	**Common name**
Gerreidae	*Gerres filamentosus*	Whipfin silver biddy
	Gerres poieti	Strongspine silver biddy
Mullidae	*Upeneus vittatus*	Striped goatfish
Sillaginidae	*Sillago sihama*	Silver sillago
Terapontidae	*Terapon jarbua*	Jarbua terapon
Scatophagidae	*Scatophagus argus*	Spotted scat
Harpadontidae	*Harpadon nehereus*	Bombay duck
Carangidae	*Alectis indicus*	Indian threadfish
	Alepes djedaba	Shrimp scad
	Alepes melanoptera	Blackfin scad
	Atule mate	Yellowtail scad
	Carangoides caeruleopinnatus	Coastal trevally
	Carangoides malabaricus	Malabar trevally
	Carangoides oblongus	Coach whip trevally
	Caranx ignobilis	Giant trevally
	Caranx para	Banded scad
	Caranx sexfasciatus	Bigeye trevally
	Decapterus russelli	Indian scad
	Scomberoides commersonnianus	Talang queenfish
	Scomberoides toli	Needle scaled queenfish
	Selaroides leptolepis	Yellowstripe scad
	Carangid sp.	*Carangid sp.*
Haemulidae	*Plectorhinchus pictus*	Trout sweet lips
	Pomadasys argyreus	Bluecheek silver grunt
	Pomadasys commersonni	Spotted Grunter
	Pomadasys kaakan	Javelin grunter
	Pomadasys maculatum	Saddle grant
	Pomadasys olivaceum	Olive grant
	Pomadasys stridens	Stripped grunter
Sparidae	*Acanthopagrus berda*	Picnic seabream
	Acanthopagrus latus	Yellowfin seabream
	Argyrops spinifer	King soldier bream
	Cheimerius nufar	Santer seabream
	Crenidens crenidens	Karanteen seabream
	Rhabdosargus sarba	Goldlined seabream

Permanent Finfishes of Sonmiani Bay		
Family	**Species**	**Common name**
Sciaenidae	*Johnieops sina*	Sin croaker
	Johnius aneus	Greyfin croaker
	Johnius belangerii	Belanger's croaker
	Johnius carutta	Karut croaker
	Johnius dussumieri	Bearded croaker
	Johnius glaucus	Pale spotfin croaker
	Otolithes cuvieri	Lesser tiger tooth croaker
	Otolithes ruber	Tiger tooth croaker
	Paranibea semiluctuosa	Half mourning croaker
	Umbrina canariensis	Canary drum
Lethrinidae	*Lethrinus microdon*	Smalltooth emperor
	Lethrinus ramak	Yellow banded emperor
Serranidae	*Epinephelus fuscoguttatus*	Brown-marbled grouper
Muraenesocidae	*Muraenesox bogio*	Common pike conger
Ophichthidae	*Cirrhimuraena playfairii*	Fringe lip snake eel
Hemiscylliidae	*Chiloscyllium griseum*	Gray bamboo shark
Narcinidae	*Narcine indica*	Large spotted numbfish
Narcinidae	*Narcine timlei*	Black spotted electric ray
Gymnuridae	*Gymnura poecilura*	Butterfly-ray
Pelagic waters dweller finfishes of Sonmiani Bay		
Family	Species	Common name
Ambassidae	*Ambasis gymnocephalus*	Bald Glassy Perchlet
	Ambassis ambassis	Commerson's glassy
Belonidae	*Strongylura strongylura*	Spottail needlefish
	Tylosurus crocodiles crocodilus	Crocodile needlefish
Chirocentridae	*Chirocentrus nudus*	Whitefin wolf-herring
Clupeidae	*Dussumieria acuta*	Rainbow sardine
	Escualosa thoracata	White sardine
	Hilsa kelee	Kelee shad
	Ilisha megaloptera	Bigeye ilisha
	Ilisha melastoma	Indian ilisha
	Opisthopterus tardoore	Tardoore
	Sardinella albella	White sardinella
	Sardinella gibbosa	Gold stripe sardinella
	Sardinella longiceps	Oil sardine
	Sardinella melanura	Blacktip sardinella
	Sardinella sindensis	Sind sardinella
	Tenualosa toli	Toli shad
	Ilisha sp.	*Ilisha sp.*

Permanent Finfishes of Sonmiani Bay		
Family	**Species**	**Common name**
Engraulidae	*Coilia dussumieri*	Gold spotted grenadier anchovy
	Stolephorus commersonii	Commerson's anchovy
	Stolephorus indicus	Indian anchovy
	Thryssa hamiltonii	Hamilton's thryssa
	Thryssa malabarica	Malabar thryssa
	Thryssa mystax	Mustached thryssa
	Thryssa setirostris	Long jaw thryssa
	Thryssa vitrirostris	Orange mouth anchovy
	Thryssa sp.	*Thryssa sp.*
Hemiramphidae	*Hyporhamphus dussumieri*	Dussumier's halfbeak
	Hyporhamphus limbatus	Congaturi halfbeak
Sphyraenidae	*Sphyraena obtusata*	Obtuse barracuda
Lactariidae	*Lactarius lactarius*	False trevally
Carangidae	*Megalaspis cordyla*	Torpedo scad
	Parastromateus niger	Black pomphret
Drepanidae	*Drepane punctata*	Spotted sickle fish
Nemipteridae	*Nemipterus bipunctatus*	Delagoa threadfin bream
Trichiuridae	*Lepturacanthus savala*	Savalai hairtail
	Trichiurus lepturus	Largehead hairtail
Scombridae	*Auxis thazard*	Frigate tuna
	Rastrelliger kanagurta	Indian mackerel
Stromateidae	*Pampus argenteus*	Silver pomfret
	Pampus chinensis	Chinese silver pomfret
Aluteridae	*Alutera monoceros*	Yellow finned leather jacket
Tetraodontidae	*Arothron lepoardus*	Banded leopard blowfish
	Lagocephalus laevigatus	Smooth puffer
	Lagocephalus lunaris	Green rough backed blowfish
Triacanthidae	*Pseudotriacanthus strigilifer*	Long-spined tripod fish

Table 3.
Permanent and seasonal diversity of Finfishes in two habitat of the Sonmiani Bay, during the study period, December 2001 to November 2002.

fins and moist outer coat. The pectoral fins have bases to help them creep through the mud with more vigor and moist outer coat acts as a breathing organ respectively.

3.1.2 Seasonal finfishes of Sonmiani Bay

Seasonal finfishes in the area include the family Bothidae, which includes the species *Pseudorhombus elevates,* and the Soleidae, which includes the species *Solea elongate,* as revealed in our findings (**Table 3**).

3.1.3 Demersal waters dweller finfishes of Sonmiani Bay

Ariidae and Lutjanidae [134] are mostly demersal finfishes which usually found on the soft bentho-pelagic area and mostly travel toward the high tidal zone. (**Table 3**). Flatfish, flathead, rays, and demersal finfishes are buried in the mud of mangrove creeks, and while giant fish may live in deep waters, lesser species access the Sonmiani bay's tidal channels.

3.1.4 Pelagic waters dweller finfishes of Sonmiani Bay

Pelagic water dweller finfishes are those that only stay for a short time during tidal intervals, leaving at low tide and returning at high tide (**Table 3**). The family Engraulidae and clupeids are belonging to this group which are plankton feeder such as sardine, shad, and herring fishes. These fishes are fast swimmers.

3.2 Carnivorous and herbivorous finfishes of the Sonmiani Bay

The collected specimens of finfishes throughout the study were split into two groups based on their mode of nutrition; Bianchi [91] was followed during the observation. Fish were divided into groups based on their feeding habits and food preferences (**Table 4**). The finfishes of Sonmiani Bay were divided in this way to better understand the area's food webs. **Table 4** shows the list of herbivorous and partly herbivorous fishes which are 31 species, belong to three families of bony fishes. Fishes like mullets, herring, shads, sardine, and thryssa are herbivorous while some members of family clupeidae are partly zooplankton eaters.

	Carnivorous Finfishes	
Family	**Species**	**Common name**
Batrachoididae	*Allenbatrachus grunniens*	Grunting toadfish
Bothidae	*Pseudorhombus elevatus*	Deep flounder
Soleidae	*Euryglossa orientalis*	Oriental sole
	Solea elongata	Elongate sole
Cynoglossidae	*Cynoglossus bilineatus*	Roundhead tongue sole
	Cynoglossus dispar	Carrot tongue sole
	Cynoglossus dubius	Carrot tongue sole
	Cynoglossus puncticeps	Speckled tongue sole
	Paraplagusia bilineata	Double lined tongue sole
	Cynoglossus sp.	Tongue sole
Platycephalidae	*Cociella crocodila*	Crocodile flathead
	Platycephalus indicus	Bartail flathead
Callionymidae	*Callionymus marleyi*	Sand dragonet
Synodontidae	*Sauridaundo squamis*	Brushtooth lizardfish
Polynemidae	*Eleutheronematetra dactylum*	Fourfinger threadfin
	Polynemus heptadactylus	Seven thread tassel fish
	Polynemus indicus	Indian threadfin

	Carnivorous Finfishes	
Family	**Species**	**Common name**
Ariidae	*Arius arius*	Threadfin sea catfish
	Arius caelatus	Engraved catfish
	Arius crossocheilus	Roughback sea catfish
	Arius dussumieri	Blacktip sea catfish
	Arius maculatus	Spotted catfish
	Arius platystomus	Flatmouth sea catfish
	Arius sagor	Sagor sea catfish
	Arius sona	Sona sea catfish
	Arius subrostratus	Shovelnose sea catfish
	Arius tenuispinis	Thinspine catfish
	Arius thalassinus	Giant catfish
Plotosidae	*Plotosuslineatus*	Striped eel catfish
Lutjanidae	*Lutjanusrussellii*	Russell's snapper
	Lutjanus lunulatus	Lunartail snapper
Leiognathidae	*Gazzaminuta*	Toothpony
	Leiognathus berbis	Berber ponyfish
	Leiognathus blochi	Tow-blotch ponyfish
	Leiognathus daura	Goldstripe ponyfish
	Leiognathus equulus	Common ponyfish
	Leiognathus fasciatus	Striped ponyfish
	Leiognathus splendens	splendid ponyfish
	Secutor insidiator	Pugnose ponyfish
Gerreidae	*Gerres filamentosus*	Whipfin silver biddy
	Gerres poieti	Strongspine silver biddy
Mullidae	*Upeneus vittatus*	Striped goatfish
Sillaginidae	*Sillago sihama*	Silver sillago
Terapontidae	*Terapon jarbua*	Jarbua terapon
Scatophagidae	*Scatophagus argus*	Spotted scat
Harpadontidae	*Harpadon nehereus*	Bombay duck
Carangidae	*Alectis indicus*	Indian threadfish
	Alepes djedaba	Shrimp scad
	Alepes melanoptera	Blackfin scad
	Atule mate	yellowtail scad
	Carangoides caeruleopinnatus	Coastal trevally
	Carangoides malabaricus	Malabar trevally
	Carangoides oblongus	Coach whip trevally
	Caranxig nobilis	Giant trevally
	Caranx para	Banded scad
	Caranxsex fasciatus	Bigeye trevally
	Decapterus russelli	Indian scad
	Scomberoides commersonnianus	Talang queenfish
	Scomberoides toli	Needlescaled queenfish
	Selaroides leptolepis	Yellowstripe scad
	Carangid sp.	*Carangid sp.*

	Carnivorous Finfishes	
Family	**Species**	**Common name**
Haemulidae	*Plectorhinchus pictus*	Trout sweetlips
	Pomadasys argyreus	Bluecheek silver grunt
	Pomadasys commersonni	Spotted Grunter
	Pomadasys kaakan	Javelin grunter
	Pomadasys maculatum	Saddle grant
	Pomadasys olivaceum	Olive grant
	Pomadasys stridens	Stripped grunter
Sparidae	*Acanthopagrus berda*	Picnic seabream
	Acanthopagrus latus	Yellowfin seabream
	Argyrops spinifer	King soldier bream
	Cheimerius nufar	Santer seabream
	Crenidens crenidens	Karanteen seabream
	Rhabdosargus sarba	Goldlined seabream
Sciaenidae	*Johnieops sina*	Sin croaker
	Johnius aneus	Greyfin croaker
	Johnius belangerii	Belanger's croaker
	Johnius carutta	Karut croaker
	Johnius dussumieri	Bearded croaker
	Johnius glaucus	Pale spotfin croaker
	Otolithes cuvieri	Lesser tigertooth croaker
	Otolithes ruber	Tigertooth croaker
	Paranibea semiluctuosa	Half mourning croaker
	Umbrina canariensis	Canary drum
Lethrinidae	*Lethrinus microdon*	Smalltooth emperor
	Lethrinus ramak	Yellow banded emperor
Gobiidae	*Bolephthalmus sp.*	Goby
	Oligolepis acutipennis	Sharptail goby
	Oplopomus oplopomus	Spinecheek goby
	Oxyurichthys microlepis	Maned goby
	Oxyurichthys papuensis	Frogface goby
	Trypauchen vagina	Burrowing goby
Serranidae	*Epinephelus fuscoguttatus*	Brown-marbled grouper
Muraenesocidae	*Muraenesox bogio*	Common pike conger
Ophichthidae	*Cirrhimuraena playfairii*	Fringelip snake eel
Ambassidae	*Ambasis gymnocephalus*	Bald Glassy Perchlet
	Ambassis ambassis	Commerson's glassy

	Carnivorous Finfishes	
Family	**Species**	**Common name**
Belonidae	*Strongylura strongylura*	Spottail needlefish
	Tylosurus crocodile scrocodilus	Crocodile needlefish
Chirocentridae	*Chirocentrus nudus*	Whitefin wolf-herring
Hemiramphidae	*Hyporhamphus dussumieri*	Dussumier's halfbeak
	Hyporhamphus limbatus	Congaturi halfbeak
Sphyraenidae	*Sphyraena obtusata*	Obtuse barracuda
Lactariidae	*Lactarius lactarius*	False trevally
Carangidae	*Megalaspis cordyla*	Torpedo scad
	Parastromateus niger	Black pomfret
Drepanidae	*Drepane punctata*	Spotted sicklefish
Nemipteridae	*Nemipterus bipunctatus*	Delagoa threadfin bream
Trichiuridae	*Lepturacanthus savala*	Savalai hairtail
	Trichiurus lepturus	Largehead hairtail
Scombridae	*Auxist hazard*	Frigate tuna
	Rastrelliger kanagurta	Indian mackerel
Stromateidae	*Pampus argenteus*	Silver pomfret
	Pampus chinensis	Chinese silver pomfret
Aluteridae	*Alutera monoceros*	Yellow finned leather jacket
Tetraodontidae	*Arothron lepoardus*	Banded leopard blowfish
	Lagocephalus laevigatus	Smooth puffer
	Lagocephalus lunaris	Green rough backed blowfish
Triacanthidae	*Pseudotriacanthus strigilifer*	Long-spined tripodfish
	Carnivorous Elasmobranch fishes	
Dasyatidae	*Himantura gerrardi*	Sting ray
Narcinidae	*Narcine indica*	Large spotted numb fish
Narcinidae	*Narcine timlei*	Black spotted electric ray
Gymnuridae	*Gymnura poecilura*	Butterfly-ray
Hemiscylliidae	*Chiloscyllium griseum*	Gray bamboo shark
	Herbivorous finfishes	
Family	**Species**	**Common name**
Mugillidae	*Liza carinata*	Keeled mullet
	Liza subviridis	Greenback mullet
	Liza tada	Tade gray mullet
	Valamugil cunnesius	Longarm mullet
	Valamugil seheli	Blue spot mullet
	Valamugil speigleri	Speigler's mullet
	Liza sp.	Liza Juvenile

	Carnivorous Finfishes	
Family	**Species**	**Common name**
Clupeidae*	*Dussumieria acuta*	Rainbow sardine
	Escualosa thoracata	White sardine
	Hilsa kelee	Kelee shad
	Ilisha megaloptera	Bigeye ilisha
	Ilisha melastoma	Indian ilisha
	Opisthopterus tardoore	Tardoore
	Sardinella albella	White sardinella
	Sardinella gibbosa	Goldstripe sardinella
	Sardinella longiceps	Oil sardine
	Sardinella melanura	Blacktip sardinella
	Sardinella sindensis	Sind sardinella
	Tenualosa toli	Toli shad
	Anodontostoma chacunda	Chacunda gizzard shad
	Nematolosa nasus	Bloch's gizzard shad
	Ilisha sp.	*Ilisha sp.*
Engraulidae	*Coilia dussumieri*	Gold spotted grenadier anchovy
	Stolephorus commersonii	Commerson's anchovy
	Stolephorus indicus	Indian anchovy
	Thryssa hamiltonii	Hamilton's thryssa
	Thryssa malabarica	Malabar thryssa
	Thryssa mystax	Mustached thryssa
	Thryssa setirostris	Longjaw thryssa
	Thryssa vitrirostris	Orangemouth anchovy
	Thryssa sp.	*Thryssa sp.*
*Partly zooplankton eaters		

Table 4.
Carnivorous and herbivorous finfishes collected in the Sonmiani Bay, Balochistan during the study period December 2001 to November 2002.

There were 119 species of bony fish from 42 families and 5 species of elasmobranches from four families among the carnivorous finfishes (**Table 4**). Sharks and rays dwell in the sea by nature, although some live in estuaries and bays, where they are voracious feeders among the school of fish. They are predators with formidable jaws that attack their prey. In the Sonmiani Bay waters, carnivorous fishes were discovered to be far more abundant than herbivorous fishes (**Table 4**).

4. Discussion

Fish feeding habits and trophic relationships always remain in focus and interest in scientific essays for ages therefore, fish diets have been extensively studied, primarily through stomach content analysis, as providing important information on the ecology, physiology, and ethology of species, with huge difference in the way applications [135, 136].

Coastal waters especially lagoons have great importance because they provide feeding and nursery ground to a variety of fish and other species, highly dynamic because of high temperature and high transmission of light up to the bottom [137]. These water bodies can be exposed to very strong fluctuations of salinities due to seasonal variations of precipitation and evaporation. In addition, in the shallow environments, the seasonal fluctuations of temperatures are often more pronounced than in the adjacent sea and can make stressing conditions for many aquatic species. Due to such variabilities, their vulnerability makes these water bodies as a unique position at terrestrial, freshwater, and marine interfaces [138]. Coastal lagoons, along with estuarine environments and coastal wetlands, have been defined as Critical Transition Zones (CTZs). Those organisms that cover these areas as distributed in order to three-dimensional scales that are more likely to subordinate with the environment and reserve consumption [139]. The biotic and abiotic factors support creation of numerous types of niches and habitats for living and dispersal of organisms [140]. The temperature and salinity may cause the divergence in structure and function of many benthic and pelagic communities because these communities are constantly interacting and suffering from these ecological factors [141–144]. The higher temperature supports the productivity and ultimately abundance of species in particular areas [145–147]. The feeding habits of the fauna may change due to the effect of salinity so therefore, salinity has a great influence on ecosystem as well as distribution and feeding habits of residential species. Tidal fluctuations cause a significant change and play a vital role in communication and other approaches which juveniles and other fauna adopted to take a sequential change [148]. These are the areas which organisms utilize and are distinct in habitat differences and accessibility due to the tidal flux [149]. However, seasonal and monthly variations also affect intertidal shallow environments, productivity, and the distribution and diversity of fish species.

In spite of primary productivity, the presence of vegetation also improves the food opportunity to the herbivores and carnivores species as the mangrove vegetation and allied mudflats have been reported as an important breeding and feeding site of a number of marine species [15]. The ecological status and environmental conditions of habitat can be the indicator of the faunal diversity and also provide the detailed information about the life history of fish species and decapods, and supportive feeding environment along with habitat requirements to complete the lifecycle are supposed to be helpful in the protection and restoration of the communities inhabiting in these areas [6, 150].

The marine fisheries policies of developing countries such as Pakistan are aimed at achieving the following objectives: filling protein gaps as regards improving marine fish supplies for domestic use, encouraging jobs, growing fishermen's economic interests, and increasing foreign exchange through exporting fish and shellfish. Maximum attention is being paid in Pakistan to achieve the ultimate goal that is, earning foreign exchange, which has established the marine fishing industry of Pakistan as an export supplier. The need for protection of fishery resources arises from the industrial sector, as it is understood from the natural predation of fish stocks [151]. Food quality and quantity are the two biggest exogenous elements impacting fish growth and indirectly, maturation and mortality, thus linked to health of fish [135]. Traditionally, data on the quality and quantity of food consumed by fish, which can be derived from the feeding habit studies, which has just been made available for fisheries research by incorporating it into appropriate fisheries models likewise; multispecies virtual population analysis which, after scaling up to the overall biomass of predators and prey, provides estimates of the total biomass consumed by predators [135]. There are numerous challenges that need to be studied for the feeding biology of particular species as also variable in a group of same species; developmental or growth diversity includes morphological diversity from the smallest to the largest in body size; as the species passes through various ontogenic stages during its development and may have a preference of different types of food during each stage therefore, exhibit variable feeding habits; behavioral diversity due to exploration of high habitat diversity expanding from marine to freshwater due to migratory in nature, etc. Few scientists have studied the diet variations in fishes and explain the changes in diet composition with reference to habitat and season [152]. Whereas, the different authors have worked on feeding habits of various finfishes [134, 153–155]. It has been reported that 25 different food items represent zooplankton (0.54%), phytoplankton (82.53%), algae (0.92%), copepod fragment (2.69%), debris (4.86%), plant-like matter (7.34%), and unidentified matters (0.77%) in some clupediae species [156]. Investigation on feeding habitat of Sin croaker narrated that it's active carnivore and depends on the benthic crustacean, fishes, shrimp as food [156].

Systematic analysis of feeding characteristics of marine fish species during the early development stage, feeding habits, and growth of larvae and juveniles can provide the information about the nutrient feeding, that should be given through biological food for nutritional improvement and enhance the survival rate of fingerlings in culture condition. Therefore, systematic study serves as a model for large-scale generation of larvae and juveniles of marine fish [156].

Not only the opportunistic feeders, some other species also obtain their food according to the availability and abundance in their habitat and like opportunistic feeders, they adapted different ways to obtain food such as basic food, which is only utilized in favorable conditions, second is incidental food that's utilized during unfavorable conditions, and then obligatory food that comprises anything that is fed for survival not as a habit. A "food-based conservation" approach is favored with the eco-reinstatement as deal with the fish habitat and fish communities. It seems extremely obvious now not to rely on abstract sampling procedures for merely descriptive assessment of food and feeding biology in fish. Therefore, investigating only the ingested food to examine the feeding habit of fish is not a reliable method and items present in the gut addressed the single picture of feeding habit because feeding habit depends upon the availability of food in the environment. So, this method just provides a piece of limited information and does not explain the factors affecting the gaining or selection of food. It has been suggested to collect the plankton community when fish is sampled for the study so, it can be helpful to determine the availability and selectivity of food [95]. Food science is an applied science and the detailed feeding biology of fish can contribute to formulate feeding design for better management and growth of fish for the culture of species. Due to the contradiction and insufficient information about available food and feeding habit, molecular level studies have been suggested to validate the information on feeding habits as obtaining of food through smelling is related with some chemicals in the fish body. Therefore, the study through molecular taste science can also provide advanced tools and techniques in biological sciences.

5. Conclusions

Lagoons are the imperative feature of any coast line that offers important ecosystem services, such as coastline protection, shelter and food for migratory and resident animals, fisheries resources, and recreation for human populations. These intertidal benthic and pelagic areas have a variety of sizes according to species diversity and distribution of species is greatly influenced by the change in biotic and abiotic factors; Abiotic factors mainly include salinity and temperature as affect the intertidal shallow environments, primary productivity, and the distribution and diversity of fish species. The ecological status and habitat environmental conditions can be the indicator of the faunal diversity and also provide the detailed information about the life history of fish species and deca-pods, along with particular supportive feeding environment and habitat requirements to complete the life cycle. These ecological indicators supposed to be helpful in the protection and restoration of the communities inhabiting in these areas. There are numerous challenges that need to be studied for the feeding biology of particular species as also variable in a group of same species and as well as developmental or growth feeding diversity as the species passes through various ontogenic stages during its development and may have a preference of different types of food during each stage therefore, exhibit variable feeding habits; behavioral diversity due to exploration of high habitat diversity.

Conflict of interest

The authors declare no conflict of interest.

Author details

Noor Us Saher[1*], Raoof M. Niazi[2], Altaf Hussain Narejo[1], Noor Hawa[1], Abdul Hameed Baloch[1], Muhammed Tabish[1], Mussarat ul Ain[1], Faiqa Razi[1] and Naureen Aziz Qureshi[3]

1 Centre of Excellence in Marine Biology, University of Karachi, Pakistan

2 Govt. Degree Boys Science College 5L New Karachi, Pakistan

3 Department of Zoology, Govt College University, Faisalabad, Pakistan

*Address all correspondence to: noorusaher@gmail.com

References

[1] Dawson A, Palmer TN, Corti S. Simulating regime structures in weather and climate prediction models. Geophysical Research Letters. 2012;**39**(21)

[2] Dawson TP, The UK. Government agrees to create the world's largest marine reserve around the Pitcairn Islands, a UK overseas territory in the South Pacific. Pacific Conservation Biology. 2015;**21**:108-112. DOI: 10.1071/PC14916

[3] Mahesh V, Benakappa S, Dineshbabu AP, Naik AK, Vijaykumar ME, Khavi M. Occurrence of low value bycatch in trawl fisheries off Karnataka, India. Fishery Technology. 2017;**54**(4):227-236

[4] Brown AC, McLachlan A. Ecology of Sandy Shores. Elsevier; 1990

[5] Fortes WLS, Almeida-Silva PH, Prestrelo L, Monteiro-Neto C. Patterns of fish and crustacean community structure in a coastal lagoon system, Rio de Janeiro, Brazil. Marine Biology Research. 2014;**10**(2):111-122. DOI: 10.1080/17451000.2013.797645

[6] Beck MW, Heck KL, Able KW, Childers DL, Eggleston DB, Gillanders BM, et al. The identification, conservation, and management of estuarine and marine nurseries for fish and invertebrates. Bioscience. 2001;**51**:633-641

[7] Rotherham D, Underwood AJ, Chapman MG. Gray CA.A strategy for developing scientific sampling tools for fishery-independent surveys of estuarine fish in New South Wales, Australia. ICES Journal of Marine Science. 2007;**64**:1512-1516

[8] Sheaves M, Johnston R. Influence of marine and freshwater connectivity on the dynamics of subtropical estuarine wetland fish metapopulations. Marine Ecology Progress Series. 2008;**537**:225-243

[9] Kjerfve B. Coastal Lagoon. In: Kjerfve B, editor. Coastal Lagoon Processes. Amsterdam: Elsevier Oceanogr. Ser; 1994. pp. 1-8

[10] Carlos EGF, Kintz JRC, Ladino LMML, Cortes FA. Benthic macrofauna associated with submerged bottoms of a tectonic estuary in tropical Eastern Pacific. Journal of Marine Biology. 2011;**2011**:1-13

[11] SteinIII W, Smith PW, Smith G. The cast net: An overlooked sampling gear. Marine and Coastal Fisheries. 2014;**6**(1):12-19

[12] Poyyamoli G, Padmavathy A, Balachandran N. Coastal sand dunes - vegetation structure, diversity and disturbance in Nallavadu Village (Puducherry, India). Asian Journal of Air Water Environment. 2011;**8**:115-122

[13] Nichols MM, Boon JD. Sediment transport processes in coastal lagoons. In: Kjerfve B, editor. Coastal lagoon processes. Vol. 7. Amsterdam, The Netherlands: Elsevier; 1994. pp. 157-219

[14] Monkolprasit S. Fish in Mangroves and Adjacent Areas. In: Paper presented in the First UNDP/UNESCO Regional Training Course on Introduction to Mangrove Ecosystems Held from 2-30 March. Bangkok, Thailand; 1983

[15] Gondal MA, Saher NU, Qureshi NA. Diversity and biomass distribution of intertidal Fauna in Sonmiani Bay (MianiHor), Balochistan (Pakistan). Egyptian Academic Journal of Biological Sciences. 2012;**4**:219-234

[16] Blaber SJM, Young JW, Dunning MC. Community structure and zoogeographic affinities of the coastal fishes of the Dampier region of North-Western Australia. Marine and Freshwater Research. 1985;**36**(2):247-266

[17] Ley JA, McIvor CC, Montague CL. Fishes in mangrove prop-root habitats of northeastern Florida Bay: Distinct assemblages across an estuarine gradient. Estuarine, Coastal and Shelf Science. 1999;**48**(6):701-723

[18] Ikejima K, Tongnunui P, Medej T, Taniuchi T. Juvenile and small fishes in a mangrove estuary in Trang province, Thailand: Seasonal and habitat differences. Estuarine, Coastal and Shelf Science. 2003;**56**(3-4):447-457

[19] Amezcua F, Amezcua-Linares F. Seasonal changes of fish assemblages in a subtropical lagoon in the SE Gulf of California. The Scientific World Journal. 2014;**2014**:15. Article ID 968902. DOI: 10.1155/2014/968902

[20] Ahmed M, Abbas G. Abundance of the finfish and shell fish juveniles in the intertidal zone of MianiHor lagoon in Balochistan, Pakistan. Pakistan Journal of Zoology. 1999;**31**:187-195

[21] Gondal MA. Comparative study of community structures of benthic macrofauna in Sonmiani Bay (MianiHor) Balochistan, Pakistan. (Ph.D Thesis). Center of Excellence. Marine Biology. 2013

[22] Ahmed M, Ayub Z. First report on the penaeid shrimp population from MianiHor lagoon in Balochistan, Pakistan. Pakistan Journal of Zoology. 1996;**28**:331-334

[23] Rodelli MR, Gearing JN, Gearing PJ, Marshall N, Sasekumar A. Stable isotope ratios as a tracer of mangrove carbon in Malaysian ecosystems. Oecologia. 1984;**61**:326-333

[24] Sheaves M, Molony B. Short-circuit in the mangrove food chain. Marine Ecology Progress Series. 2000;**199**:97-109

[25] Warlen S, Able KW, Laban E. Recruitment of larval Atlantic menhaden (Brevoortiatyrannus) to North Carolina and New Jersey estuaries: evidence for larval transport northward along the east coast of the United States. Fish Bulletin (Wash DC). 2002; 100:609-623.

[26] Lopez-Peralta RH, Arcila CAT. Diet Composition of Fish Species from the Southern Continental Shelf of Colombia. 2002.

[27] Freire KM, Christensen V, Pauly D. Description of the East Brazil large marine ecosystem using a trophic model. Scientia Marina. 2008;**72**(3):477-491

[28] Lott JP, Willis DW, Lucchesi DO. Relationship of food habits to yellow perch growth and population structure in South Dakota lakes. Journal of Freshwater Ecology. 1996;**11**(1):27-37

[29] Peyami FY. Food and feeding habit of a Teleostean fish Salmostomabacaila (Ham.) at Partapur dam, MakhdumpurJehanabad, Bihar. International Journal of Fisheries and Aquatic Studies. 2018;**6**(4):388-391

[30] Kulbicki M, Guillemot N, Amand M. A general approach to length-weight relationships for New Caledonian lagoon fishes. Cybium. 2005;**29**(3):235-252

[31] Wotton RJ. Ecology of Teleost Fishes. Fish and Fisheries Series 1. London: Chapman & Hall; 1990, 1990. p. 404

[32] Bolnick DI, Svanbäck R, Fordyce JA, Yang LH, Davis JM, Hulsey CD, et al. The ecology of individuals: Incidence and implications of individual specialization. The American Naturalist. 2003;**161**(1):1-28

[33] Sivan G, Radhakrishnan CK. Food, feeding habits and biochemical composition of Scatophagus argus. Turkish Journal of Fisheries and Aquatic Sciences. 2011;**11**(4):603-608

[34] Karpouzi VS, Stergiou KI. The relationships between mouth size and shape and body length for 18 species of marine fishes and their trophic implications. Journal of Fish Biology. 2003;**62**(6):1353-1365

[35] Naranjo-Elizondo B, Espinoza M, Herrera M, Clarke TM, Wehrtmann IS. Feeding habits of the Pacific bearded brotula Brotula clarkae Hubbs, 1944 (Ophidiidae) along the Pacific coast of Costa Rica, Central America. Journal of Applied Ichthyology. 2016:1-9. DOI: 10.1111/jai.13029

[36] Allan JD, Castillo MM. Stream Ecology: Structure and Function of Running Waters. 2nd ed. Dordrecht, The Netherlands: Springer; 2007

[37] Pelicice FM, Agostinho AA. Feeding ecology of fishes associated with Egeria sps. Patches in a tropical reservoir, Brazil. Ecology of Freshwater Fish. 2006;**15**:10-19

[38] Baker R, Buckland A, Sheaves M. Fish gut content analysis: Robust measure of diet composition. Fish and Fisheries. 2014;**15**:170-177

[39] Ellis JR, Pawson MG, Shackley SE. The comparative feeding ecology of sixspecies of shark and four species of ray (Elasmobranchii) in the north-East Atlantic. Journal of the Marine Biological Association. 1996;**76**:89-106

[40] Pinkas L, Oliphant MS, Iverson ILK. Food habits of albacore, bluefin tuna, and bonito in California waters. California Department of Fisheries Game. Fisheries Bulletin. 1971;**152**:105

[41] Hyslop EJ. Stomach content analysis – A review of methods and their application. Journal of Fish Biology. 1980;**17**:411-429

[42] Kawakami E, Vazzoler G. Método gráfico e estimativa de índice alimentar aplicado no estudo de alimentação de peixes. Boletim do Instituto oceanográfico. 1980;**29**:205-207

[43] Hajisamae S, Chou LM, Ibrahim S. Feeding habits and trophic organization of the fish community in shallow waters of an impacted tropical habitat. Estuarine, Coastal and Shelf Science. 2003;**58**:89-98

[44] Figueiredo MB, Carvalho Neta RNF, Silva Nunes JL, da Silva de Almeida Z. Hábitosalimentarios de Macrodonancylodon (Actinopterygii, Sciaenidae) enelnoreste de Brasil. Revista de biología marina y oceanografía. 2014;**49**(3):559-566

[45] Pothoven SA, Nalepa TF. Feeding ecology of lake Whitefish in Lake Huron. Journal of Great Lakes Research. 2006;**32**:489-501

[46] Ivlev VS. Experimental Ecology of the Feeding of Fishes. Connecticut: Yale University Press; 1961

[47] Pianka ER. The structure of lizard communities. Annual Review of Ecology and Systematics. 1973;**4**:53-74

[48] Hulbert SH. The measurement of niche overlap and some relatives. Ecology. 1978;**59**:67-77

[49] Krebs CJ. Ecological Methodology. New York: Harpet and Row; 1989

[50] Marshall S, Elliot M. A comparison of univariate and multivariate numerical andgraphical techniques for determining inter and intraspecific feeding

relationships in estuarine fish. Journal of Fish Biology. 1997;**51**:526-545

[51] Pelicice FM, Agostinho AA, Thomaz SM. Fish assemblages associated Egera in a tropical reservoir: Investigating the effects of plant biomass and diel period. Acta Oecologia. 2005;**27**:9-16

[52] Saikia SK. Proposal of new and simple descriptive measure of diet breadth. Ecología. 2012;**2**(3):85-92

[53] Carvalho-Filho A. Peixes: Costa Brasileira. São Paulo: EditoraMelro; 1999. p. 320

[54] Morasche MS, Tubino RDA, Monteiro NC. (2010). Dieta da corvina, Micropogonias furnieri (Desmarest, 1823)(Actinopterygii, Sciaenidae) na região Costeira de Itaipu, Niterói–RJ. Ciências do Mar. 2010;**43**(2):87-95

[55] Hynes HBN. The food of fresh water sticklebacks (Gasterosteus aculeatus) and (Pygosteuspungitius) with a review of methods used in studies of food of fishes. Journal of Animal Ecology. 1950;**19**:36-58

[56] Pianka ER. Evolutionary Ecology. New York: Harper & Row; 1988

[57] Bizzarro JJ, Robinson HJ, Rinewalt CS, Ebert DA. Comparative feeding ecology of four sympatric skate species off Central California, USA. In: Biology of Skates. Dordrecht: Springer; 2007. pp. 91-114

[58] Munoz RC, Currin CA, Whitfield PE. Diet of invasive lionfish on hard bottom reefs of the Southeast USA: Insights from stomach contents and stable isotopes. Marine Ecology Progress Series. 2011;**432**:181-193

[59] Robertson DR, Pérez-España H, Domínguez-Domínguez O, Estapé CJ, Estapé AM. An update to the inventory of shore-fishes from the Parque Nacional Sistema ArrecifalVeracruzano, Veracruz, México. ZooKeys. 2019;**882**:127-157

[60] Félix PM, Amorim MCP, Pereira TJ, Fonseca PJ, Sousa-Santos C, Costa JL. Feeding ecology and life-history strategy of nesting males in a fish with long parental care, Lusitanian toadfish (Halobatrachus didactylus, Batrachoididae). Journal of the Marine Biological Association of the United Kingdom. 2016;**96**(3):657-665

[61] Ruiz L. Morfología del tracto digestivo y hábitos alimentarios de Thalassophryne maculosa (Pisces: Batrachoididae) de la bahía de Mochima, Venezuela. Boletín del centro de investigaciones biológicas. 2010;**44**:51-62

[62] Bob-Manuel FG. Food and feeding ecology of the mudskipper Periopthalmuskoelreuteri (PALLAS) Gobiidae at Rumuolumeni creek, Niger Delta, Nigeria. Agriculture and Biology Journal of North America. 2011;**2**(6):897-901

[63] Shukla ML, Trivedi JN, Soni GM, Patel BK, Vachhrajani KD. Mudskipper (Gobiidae: Oxudercinae) fauna of Northern Gulf of Khambhat with two new record of the species from Gujarat. India European Journal of Zoological Research. 2014;**3**(3):67-74

[64] Mahadevan G, Ravi V. Distribution of mudskippers in the mudflats of muthupet, southeast coast of India. International Journal of Fisheries and Aquatic Studies. 2015;**3**(2):268-272

[65] Chernoff B. Atherinidae and Atherinopsidae. In: Carpenter KE, editor. The living marine resources of the Western Central Atlantic. Volume 3: Bonyfishes part 2 (Opistognathidae to Molidae) FAO species identification

guide for fishery purposes and American Society of Ichthyologist and Herpetologists Special Publication No. 5. Rome: FAO; 2003. pp. 1086-1103

[66] Grabowska J, Grabowski M. Diel-feeding activity in early summer of racer goby Neogobiusgymnotrachelus (Gobiidae): A new invader in the Baltic basin. Journal of Applied Ichthyology. 2005;**21**(4):282-286

[67] Hesthagen IH. Migrations, breeding, and growth in Pomatoschistusminutus (Pallas)(pisces, Gobiidae) in Oslofjorden. Norway. Sarsia. 1977;**63**(1):17-26

[68] Achakzai WM, Saddozai S, Baloch WA, Massod J, Rehman HU, Ain MU. Food and feeding habits of Glossogobiusgiuris (Hamilton and Buchannan, 1822) collected from Manchar Lake distt. Jamshoro, Sindh, Pakistan. Global Veterinaria. 2015;**14**(4):613-618

[69] Rao LM, Rao PS. Food and feeding habits of Glossogobiusgiuris from Gosthani estuary. Indian Journal of Fisheries. 2002;**49**(1):35-40

[70] Dinh QM, Qin JG, Dittmann S, Tran DD. Seasonal variation of food and feeding in burrowing goby Parapocryptesserperaster (Gobiidae) at different body sizes. Ichthyological Research. 2017;**64**(2):179-189

[71] Števove B, Kováč V. Do invasive bighead goby Neogobius kessleri and round goby N. melanostomus (Teleostei, Gobiidae) compete for food? Knowledge and Management of Aquatic Ecosystems. 2013;**410**:08

[72] Esposito V, Castriota L, Consoli P, Romeo T, Falautano M, Andaloro F. Feeding habits and selectivity of the wide-eyed flounder, Bothuspodas (Delaroche, 1809)(Bothidae) from the southern Tyrrhenian Sea. Marine Biology Research. 2010;**6**(5):496-502

[73] Bayhan B, Sever TM, Taşkavak ERTAN. Age, length-weight relationships and diet composition of scaldfishArnoglossuslaterna (Walbaum, 1792)(Pisces: Bothidae) in Izmir Bay (Aegean Sea). Journal of Animal and Veterinary Advances. 2008;**7**(8):924-929

[74] Khan M, Hoda SMS. Observation on the food and feeding habits of Euryglossaorientalis (Bl. and Schn) (family: Soleidae) from Karachi Coast. Journal of Islamic Academy of Sciences. 1993;**6**(3):185-188

[75] Atabak N. Survey on natural feeding of juvenile Cynoglossusarel and Soleaelongata fishes (Cynoglossidae and Soleidae) in the northwest of Persian gulf coastal water. Fisheries Research. 2011;**108**(1):9-14

[76] Tim-IBP K. Field Guide of Fish, Marine Mammal, Sea Turtle, and Mangrove in Berau-Bintuni Bay. West Papua: InstitutPertanian Bogor BP Berau Ltd; 2015

[77] Kailola PJ. The fishes of Papua New Guinea: A revised and annotated checklist. In: Scorpaenidae to Callionymidae.Dept. of Fisheries and Marine Resources. Vol. II. Papua New Guinea: Port Moresby; 1987

[78] Hashemi SAR, Taghavi MS. Diet composition of bartail flathead (Platycephalus indicus) in northwest of Persian Gulf. World Journal of Fish and Marine Sciences. 2013;**5**(1):35-41

[79] Ramanujam ME, Anbarasan R. A priliminary report on the ichthyofauna of Yedayanthittu Estuary (Tamil Nadu, India) and rivulets draining into it. Journal of Threatened Taxa.

2009;**1**(5):287-294 www.threatenedtaxa.org | May 2009 |

[80] Bauchot ML. Poissonsosseux. In: Fischer W, Bauchot ML, Schneider M, editors. Fiches FAO d'identification pour les besoins de la pêche. (rev. 1). Méditerranée et mer Noire. Zone de pêche 37. Vol. II. Rome: Commission des CommunautésEuropéennes and FAO; 1987. pp. 891-1421

[81] Mouneimné N. Liste des poissons de la côte du Liban (Mediterranéeorientale). Cybium. 1977;**1**:37-66

[82] Allen GR, Dudgeon CL. Hemiscylliummichaeli, a new species of Bamboo Shark (Hemiscyllidae) from Papua New Guinea. International Journal of Ichthyology. 2010;**16**(1):19-30 (Ref. 83755)

[83] Afonso P, Porteiro FM, Santos RS, Barreiros JP, Worms J, Wirtz P. Coastal marine fishes of São Tomé Island (Gulf of Guinea). Arquipélago. 1999;**17**(A):65-92

[84] Nelson JS. Fishes of the World. Third ed. New York: John Wiley & Sons, Inc.; 1994. p. 600

[85] Sulak KJ. Synodontidae. In: Quero JC, Hureau JC, Karrer C, Post A, Saldanha L, editors. Check-list of the Fishes of the Eastern Tropical Atlantic (CLOFETA). Lisbon: JNICT; 1990. pp. 365-370

[86] Scott WB, Scott MG. Atlantic fishes of Canada. Canadian Bulletin Fish and Aquatic Science. 1988;**219**:731

[87] Robins CR, Ray GC. A Field Guide to Atlantic Coast Fishes of North America. USA: Houghton Mifflin Company, Boston; 1986. p. 354

[88] Allen G, Robertson D. Fishes of the Tropical Eastern Pacific. Honolulu, HI: University of Hawaii Press; 1994

[89] Wheeler A. The World Encyclopedia of Fishes. London: Macdonald; 1985

[90] Lewis AD, Pring CK. Freshwater and brackishwater fish and fisheries of Fiji. In: Petr T, editor. Reports and papers presented at the Indo-Pacific Fishery Commission Expert Consultation on inland fisheries of the larger Indo-Pacific islands. Bangkok, Thailand: FAO Fish; 1986. pp. 38-59. Rep. (371) Suppl.:258 p. 1986

[91] Bianchi G. Field Guide of Commercial Marine and Brackish Water Species of Pakistan. Rome Italy: FAO Identification Sheets for Fishery Purpose; 1985

[92] Cardona L. Effects of salinity on the habitat selection and growth performance of Mediterranean flathead grey mullet Mugil cephalus (Osteichthyes, Mugilidae). Estuarine, Coastal and Shelf Science. 2000;**50**(5):727-737

[93] Motomura H. Family Polynemidae Rafinesque 1815-threadfins. Annot. Checklists Fish., Calif. Acad. Sci., San Francisco. 2004;**32**:1-18

[94] Psomadakis PN, Osmany HB, Khan W. Important Coastal Fishery Species of Pakistan. A Pocket Guide. Rome: FAO; 2014

[95] Koslow JA. Feeding selectivity of schools of Northern Anchovy, Engraulismordax, in the Southern California bight. Fishery Bulletin. 1981;**79**(1):131-142

[96] Ali A, Siddiqui PJA, Ahmad N, Amir SA, Masroor R, Shafique S, et al. Ecology of fish communities in coral habitats along the coast of Pakistan: Potential threats and conservation strategies. Pakistan Journal of Zoology. 2021;**53**(4)

[97] Panhwar SK, Farooq N, Qamar N, Shaikh W, Mairaj M. A new Sillago species (family Sillaginidae) with descriptions of six sillaginids from the northern Arabian Sea. Marine Biodiversity. 2018;**48**(4):2225-2231

[98] Russell BC, Houston W. Offshore fishes of the Arafura Sea. The Beagle: Records of the Museums and Art Galleries of the Northern Territory. 1989;**6**:69-84

[99] Whitehead PJP, Harpadontidae. FAO species identification sheets for fishery purposes. Western Indian Ocean Fishing Area. 1984;**1984**:51

[100] Sivakami S. On the food habits of the fishes of the family carangidae a review. Journal of the Marine Biological Association of India. 1997;**38**(1 & 2):118-123

[101] Shaikh W, Panhwar SK. First record of six marine fish species caught occasionally in the northern Arabian Sea in Pakistan. Fisheries & Aquatic Life. 2021;**29**(3):158-163

[102] Iwatsuki Y. Review of the Acanthopagrus latus complex (Perciformes: Sparidae) with descriptions of three new species from the Indo-West Pacific Ocean. Journal of Fish Biology. 2013;**83**(1):64-95

[103] Amir SA, Siddiqui PJA, Masroor R. A new sparid fish of genus Sparidentex (Perciformes: Sparidae) from coastal waters of Pakistan (north western Indian Ocean). Pakistan Journal of Zoology. 2014;**46**(2):471-477

[104] Osmany HB, Zohra SK. Feeding habits of Jarbua terapon, Terapon jarbua (family Teraponidae), Forsskal from the coast of Pakistan. International Journal of Biology and Biotechnology. 2016;**13**(4):551-559

[105] Uiblein F, Heemstra PC. A taxonomic review of the Western Indian Ocean goathfishes of the genus Upeneus (Family Mullidae), with descriptions of four new species.Smithiana. Bulletin. 2010;**11**:35-71

[106] Compagno LJV, Last PR. Gymnuridae. The Living Marine Resources of the Western Pacific. 1999;**3**(part 1):1506-1510

[107] Froese R, Pauly D. FishBase world wide web electronic publications. 2019. www.fishbase.org (02/2019)

[108] Carpenter KE, Allen GR. Emperor fishes and large-eye breams of the world. FAO Fisheries Synopsis. Rome, Italy: Food and Agriculture Organization of the United Nations; 1989;**9**(125)

[109] Muhammad AA, Farooq S, Rabbaniha M, Jahangir S, Malik A, Hameed A, et al. Current fishery status of ribbonfish Trichiurus lepturus Linnaeus, 1758 (Trichiuridae) from Makran coast (northeast Arabian Sea). Iranian Journal of Fisheries Sciences. 2017;**16**(2):815-821

[110] Haedrich RL. Stromateidae. Butterfishes (harvestfishes). In: Carpenter KE, editor. FAO species identification guide for fishery purposes. The living marine resources of the Western Central Atlantic. Vol. 3: Bony fishes part 2 (Opistognathidae to Molidae), sea turtles and marine mammals. Rome Italy; 2002. pp. 1879-1884

[111] Kolandai K. Food and Feeding Habits of the Croakers,'Otolithesruber' (Schneider) Pennahiamacrophthalmus (Bleeker) and Johniusbelangerii (Cuvier), (Pisces: Sciaenidea). Laporan Projek FPSS; 1993

[112] Fonseca FAL, Castro ACL. Dinâmica da nutrição da pescada-gó, Macrodonancylodon (Bloch &Schnneider, 1801) (Teleostei,

Sciaenidae) no litoral do Maranhão. Boletim do Laboratório de Hidrobiologia. 2000;**13**:43-49

[113] Camargo M, Isaac V. Food categories reconstruction and feeding consumption estimates for the Sciaenid Macrodonancylodon(Bloch & Schneider), and the congeneric fishes Stelliferrastrifer(Jordan) and Stellifernaso(Jordan) (Pisces, Perciformes) in the Caeté Estuary, Northern Coast of Brazil. RevistaBrasileira de Zoologia. 2004;**21**:85-89

[114] WillisCM RJ, Smart T, Cowan J, Biondo P. Diet composition, feeding strategy, and diet overlap of 3 sciaenids along the southeastern United States. Fishery Bulletin. 2015;**113**(3)

[115] Santos MN, Rocha GR, Freire KM. Diet composition for three sciaenids caught off northeastern Brazil. Revista de biología marina y oceanografía. 2016;**51**(3):493-504

[116] Bandani GA, Hasanzadeh KB, Akrami R. The feeding study of tiger-toothed croaker (Otolithes rubber Bloch & Schneider, 1801) in coastal waters of Chabahar, Iran 2007

[117] Azhir MT. Biological Investigation of Tiger-Toothed Croaker, Otolithesruber, in oman Sea along Sistan and Baluchistan Province. 2008

[118] Kamali E, Dehghani R, Behzadi S, Ejlali K. A study on the some specials of biological javelin grunter (Pomadasyskaakan), tigertoothcroker (Otolithesruber) and spotted croaker (Protonibeadicanthus) in Hormozgan waters, Final report, Iranian Fisheries Research Instituted. 91P. in persian. 2006

[119] Eskandari G, Koochaknejad E, Savari A, Koochenian P, Taghavi MS. The influence of *Otolithesruber*consumption on prey and comparison with that harvested by fisheries. Journal of the Persian Gulf. 2012;**3**(9):53-61

[120] Eskandari GR. Reproduction and feedingbiology on tiger toothed croaker *Otolithesruber*inKhozestan coasts. M. S. Thesis in persian. ShahidChamran University, Faculty of Marine Science &Technology; 1997. p. 112.

[121] Pillai PK. On the biometry, food and feeding and spawning habits of Otolithesruber (Schneider) from Porto Novo. Indian Journal of Fisheries. 1983;**30**(1):69-73

[122] Renuka V, Zynudheen AA, Panda SK, Ravishankar CNR. Nutritional evaluation of processing discards from tiger tooth croaker, Otolithesruber. Food Science and Biotechnology. 2016;**25**(5):1251-1257

[123] Al-Dubakel AY. Analysis of natural food composition of fishes in Shatt Al-Arab River, southern Iraq. Jordan Journal of Biological Sciences. 2016;**9**(2):89-96

[124] Hussain NA, Naama AK, Ali TS. Feeding relationship of subtropical coastal water fish assemblage of Khor Al- Zubair, Iraq. Marina Mesopotamica. 1993;**8**(1):11-24

[125] Carpenter KE, Krupp F, Jones DA, Zajonz U. FAO Species Identification Field Guide for Fishery Purposes. Living Marine Resources of Kuwait, Eastern Saudi Arabia, Bahrain, Qatar and the United Arab Emirates. Rome: FAO; 1997.p. p. 293

[126] Akin S, Winemiller KO. Habitat use and diets of juvenile spot (Leiostomusxanthurus) and Atlantic croaker (Micropogonias undulates) in a small estuary at Mad Island marsh, Texas. Texas Journal of Science. 2015;**64**(1):1-29

[127] Taher MM, Al-Dubakel AY, Al-Lamy JH. Trophic breadth and dietary overlap for ten fish species caught from Shatt Al-Arab River, Fao, and Southern Iraq. Al- Kufa University Journal for Biology. 2016:100-109

[128] Froese R, Pauly D (Eds). FishBase. WorldWideWeb electronic publication. 2012. www.fishbase.org

[129] Madhu VR, Panda SK, Meenakumari B. Trawl Selectivity on Johnius dussumieri (Cuvier, 1830) along Gujarat, Northwest Coast of India. 2013.

[130] Shah AA, Jusoff K. Mangroves conservation through community participation in Pakistan: The case of Sonmiani Bay. International Journal of Systems applications, Engineering & Development. 2007;**4**:75-81

[131] Gondal MA, Qureshi NA, Saher NU. Morphometric Studies on Subadult Liza subviridis and Sillago sihama from Sonmiani Bay (Miani Hor), Balochistan, Pakistan. Journal of Bioresource Management. 2014;**1**(1):5

[132] Saifullah SM, Rasool F. Mangroves of MianiHor lagoon on the north Arabian Sea coast of Pakistan. Pakistan Journal of Botany. 2002;**34**(3):303-310

[133] Mukhtar I, Hannan A. Constrains on mangrove forests and conservation projects in Pakistan. Journal of Coastal Conservation. 2012;**16**(1):51-62

[134] Bachok Z, Mansor MI, Noordin RM. Diet composition and food habits of demersal and pelagic marine fishes from Terengganu waters, east coast of Peninsular Malaysia. Naga. 2004;**27**:41-47

[135] Stergiou KI, Karpouzi VS. Feeding habits and trophic levels of Mediterranean fish. Reviews in Fish Biology and Fisheries. 2002;**11**(3):217-254

[136] Karachle PK, Stergiou KI. An update on the feeding habits of fish in the Mediterranean Sea (2002-2015). Mediterranean Marine Science. 2017;**18**(1):43-52

[137] Pihl L, Rosenberg R. Production, abundance, and biomass of mobile epibenthic marine fauna in shallow waters, western Sweden. Journal of Experimental Marine Biology and Ecology. 1982;**57**(2-3):273-301

[138] Levin LA, Boesch DF, Covich A, Dahm C, Erséus C, Ewel KC, et al. The function of marine critical transition zones and the importance of sediment biodiversity. Ecosystems. 2001;**4**(5):430-451

[139] Morrison JM, Codispoti LA, Gaurin S, Jones B, Manghnani V, Zheng Z. Seasonal variation of hydrographic and nutrient fields during the US JGOFS Arabian Sea process study. Deep-Sea Research. 2002;**45**:2053

[140] Lee SY. Mangrove macro-benthos: Assemblages, services and linkages. Journal of Sea Research. 2008;**45**:40-54

[141] Day JWJ, Hall CAS, Kemp WM, Yanez-Arancibia A. Estuarine Ecology. New York: John Wiley; 1989

[142] Veloso VG, Cardoso RS, Fonseca DB. Spatio-temporal characterization of intertidal macrofauna at Prainha beach (Rio de Janeiro State). Oecol Brasil. 1997;**3**:213-225

[143] Das-Neves LP, Silva PSR, Bemvenuti CE. Temporal variability of benthic macrofauna on Cassino beach, southernmost Brazil. Iheringia, SerieZoologia. 2008;**98**:36-44

[144] Lizarralde ZI, Pittaluga S. Distribution and temporal variation of the benthic Fauna in tidal flat of

the riogallegos estuary, Patagonia, Argentina. Thalassas. 2010;**27**:9-20

[145] Gibson RN, Ansell AD, Robb L. Seasonal and annual variations in abundance and species composition of fish and macrocrustacean communities on a Scottish sandy beach. Marine Ecology Progress Series. 1993;**1983**:89-105

[146] Harris SA, Cyrus DP. Larval and juveniles fishes in the surf zone adjacent to the St. Lucia estuary mouth, KwaZulu-Natal, South Africa. Marine and Freshwater Research. 1996;**47**:465-482

[147] Layman CA. Fish assemblage structure of the shallow ocean surf-zone on the eastern shore of Virginia Barrier Islands. Estuarine, Coastal and Shelf Science. 2000;**51**(2):201-213

[148] Laroche J, Baran E, Rasoanandrasana NB. Temporal patterns in a fish assemblage of a semiarid mangrove zone in Madagascar. Journal of Fish Biology. 1997;**51**(1):3-20

[149] Wilson JP, Sheaves M. Short term temporal variations in taxonomic composition and trophic structure of a tropical estuarine fish assemblage. Marine Biology. 2001;**139**:787-196

[150] Sargent WB, Carlson PR Jr. The utility of Breder traps for sampling mangrove and high-marsh fish assemblages. In: Webb FJ, editor. Proceedings of the 14th Annual Conference on Wetlands Restoration and Creation. Tampa, FL: Hillsborough Community College; 1987. pp. 194-205

[151] Nazir K, YongtongM KMA, Memon KH, Mohsin M, Kartika S. A preliminary study on fisheries economy of Pakistan: Plan of actions for fisheries management in Pakistan. Canadian Journal of Basic and Applied Sciences. 2015;**3**(1):7-17

[152] SchaferLN PME, Valesini FJ, Potter IC. Comparisons between the influence of habitat type, season and body size on the dietary compositions of fish species in nearshore marine waters. Journal of Experimental Marine Biology and Ecology. 2002;**278**(1):67-92

[153] Yousuf AR, Bhat FA, Mehdi D, Ali S, Ahangar MA. Food and feeding habits of Ghyptosternon reticulatum (McClleland & Griffth) in torrential streams of Kashmir Himalayas. Journal of Research Development. 2003;**3**:123-133

[154] Adams CE, Woltering C, Alexander G. Epigenetic regulation of trophic morphology through feeding behaviour in Arctic charr, Salvelinus alpinus. Biological Journal of the Linnean Society. 2003;**78**(1):43-49

[155] Chrisafi P, Kaspiris P, Katselis G. Feeding habits of sand smelt (Atherinaboyeri, Risso 1810) in Tichonis Lake (Western Greece). Journal of Applied Ichthyology. 2007;**23**:209-214

[156] Suseelan C, Nair KV. Food and feeding habits of the demersal fishes off Bombay. Indian Journal of Fisheries. 1969;**16**(1&2):56-74

Section 2

Advanced (Cellular and Molecular) Approaches

Chapter 5

Sustainable Management Plans in Fisheries and Genetic Tools: An Overview of the Challenge in Invertebrates' Fisheries at the Central Area of the Southern Bay of Biscay, Spain

Marina Parrondo Lombardía, Lucía García-Florez, Eduardo Dopico Rodríguez and Yaisel Juan Borrell Pichs

Abstract

The fishing and aquaculture sectors are an important source of development around the globe. In Asturias (Spain), the diversity and richness of the fishing grounds of the Cantabrian Sea favored the historical settlement of a large number of communities closely linked to the marine environment and fishing resources, forming an integral part of the region's cultural and natural heritage. However, aquatic ecosystems are facing, nowadays, important threats from anthropogenic activities. To address these problems and avoid their impact on fishing activities, it is essential to know the ecological and genetic status of the species. Despite this, the application of genetic tools is still incipient in many species of commercial interest; however, its use can help to generate data that allow better regulation and fisheries planning. Here, the use of genetic markers and educational strategies in the management of some shellfish species of great commercial and cultural value in Asturias are reviewed. Moving toward sustainable fisheries management is a priority that can only be achieved through R + D + i, educational strategies, and the development and implementation of a regional strategy oriented toward the sustainable management and exploitation.

Keywords: small-scale fisheries, shellfish, DNA, mitochondrial DNA, microsatellites, eDNA, traceability, mislabeling, fraud, management units, connectivity, stock management, mitigation aquaculture, invasive species, game-based learning

1. Introduction

In 1984, the United Nations (UN) established an independent group of 22 individuals from member states and charged them with identifying long-term environmental strategies for the international community [1]. In the resulting report of the World Commission on Environment and Development, entitled Our Common Future—also known as the Brundtland Report [2], the term "sustainable development" was used extensively and defined as "development that meets the needs of the present without compromising the ability of future generations to meet their own needs."

The fisheries, seafood, and aquaculture sectors are an important source of food and income for millions of people around the world [3]. Addressing the problems associated with fisheries is an essential purpose, not only in the development of marine conservation policies but also for the achievement of the sustainable development goals (SDGs) of the 2030 Agenda, a major agreement that was signed in 2015 among 193 countries [4]. The achievement of these SDGs should have a strong influence on the governance of sustainable fisheries and aquaculture, ensuring that fisheries and aquaculture adapt to the impacts of climate change and improve the resilience of food production systems [3].

In addition, aquatic ecosystems face today significant threats from anthropogenic activities. In global ocean systems, concerns include climate change, overfishing, dispersal of invasive species, fertilizer runoff, plastic pollution, ocean acidification, and general defaunation [5–10]. Only 65.8% of fish stocks are currently classified as being exploited within biologically sustainable levels, continuing a downward trend that has been occurring since 1974. Similarly, underexploited species account only for 6.2% –steadily declining from 1974 to the present—whereas stocks exploited at maximum sustainable levels account for 59.6% [11].

It has been demonstrated that when fisheries are properly managed, there are significant decreases in fishing pressure and important increases in stock biomass, with some stocks having reached biologically sustainable levels, underscoring the relevant role of fisheries managers and governments when willing to take strong action [12]. The UN Code of Conduct for Responsible Fisheries states that in adopting management measures, the "best available scientific data should be used to assess the state of fishery resources" [13]. However, most exploited stocks globally are classified as data-poor stocks [14] and their status, although highly uncertain, is generally considered to be worse than that of data-rich stocks [15]. Recently, it has been argued that stock estimates based primarily on historical catch series performed on average 25% better than a random estimate; but in turn, these methods assigned fisheries the wrong FAO status category 57% of the time [16]. Substantial improvements in estimates of the state of exploited stocks worldwide depend on the expansion of new information and efficient use of existing data [16].

The use of molecular genetic techniques in fisheries research has increased dramatically in recent decades, paralleling the awareness of the value of genetic data and mainly due to the increased number of techniques available and improvements in computer technology [17, 18]. However, the application of genetic techniques to invertebrate fisheries or related problems has been remarkably scarce. Thus, most of the invertebrate groups of fishery interest have been the subject of little or no genetic study in relation to these fisheries [19].

We reviewed here the use of genetic markers as well as educative strategies in the fisheries management of some shellfish species with great commercial and cultural value in Asturias, the central area of the southern Bay of Biscay, Spain, to move forward with the relevant aim of generating data to support the design of sustainable fisheries management plans.

2. Fisheries in the central Cantabrian Sea: evidence, needs, and actions aimed at reaching sustainable exploitation levels

2.1 Fisheries in the principality of Asturias

The Principality of Asturias is an autonomous community located in northern Spain (SW Europe) bordered by the Cantabrian Sea to the north and the autonomous regions of Castilla y León, Cantabria, and Galicia to the south, east, and west, respectively. The Cantabrian Sea is the transition from the Atlantic Ocean to the Bay of Biscay, between Spain and France. The coast of Asturias covers about 30% of the Cantabrian coast and presents a general E-W trend along approximately 335 km. In general, this coastline is eminently rocky and abrupt—with a predominance of north-facing cliffs, dotted with small coves, beaches, and dune systems associated with the wider beaches [20].

Currently, the coast of Asturias is one of the most populated areas in the region, which is linked to the presence of industrial activity, such as fishing industry, factories, and ports, and to the tourist exploitation of natural resources. The historical settlement of a large number of coastal communities strongly linked to the marine environment and fishing resources was favored by the diversity and richness of the Cantabrian fishing grounds [20, 21]. These regional fishing grounds have been recently redefined and mapped, totaling 226 and occupying 984,038 hectares.

Fishing is a traditional activity linked to the Asturian coast since prehistoric times, as evidenced by archeological excavations in which numerous mollusks, crustaceans, salmonids, and other marine remains have been found [22–27]. Nineteen fishing seaports currently operate along the entire coast, and all of them are strategically located taking advantage of the sheltered location of the cliffs. These ports differ in terms of total landings, species marketed, and number of vessels. Of the total number of fishing ports, 17 of them record sales for at least 6 months of the year [21]. The National Strategic Plan of the European Fisheries Fund established that of the total number of councils that form the Asturian coastal belt, only Muros, Castrillón, and Caravia do not depend on fishing activity [28].

In Asturias, specialists agree in considering artisanal vessels registered in the category of the national census of the operational fishing fleet of "minor gears" [29], and, in 2010 there were 233 vessels registered in that category. However, according to the data collected in the Census of the Fishing Fleet registered in Asturias, in 2016 the number had decreased to 202 and, although at present (latest data from 2021) the total number of artisanal vessels totals 192—out of a total of 248 active vessels—distributed in 19 ports—representing 77.4% of the total fishing fleet with home port in Asturias –, the downward trend in the fleet of minor gears is not so pronounced, even stabilizing.

In terms of fishing strategies, most of this regional artisanal fleet switches between different gears throughout the year and exploits areas that can be reached in a few hours from fishing ports. Although the landings of the artisanal fleet—in kilograms—only represent around 10% of the total regional catch, they account for approximately 30% of the total economic value generated by the landings, due to the higher first sale price of these catches (high-value seafood products) compared to those of other fleet segments—average 4.25 €/kg compared to 1.68 €/kg –. The regional artisanal fleet targets many species, including high-value species, and landings are characterized by quality, freshness, and higher first-sale profit. This partly compensates for the lower weight of landings and lower fishing power [21, 30].

The main species landed by the Asturian artisanal fleet are hake (*Merluccius merluccius*), mackerel (*Scomber scombrus*), conger eel (*Conger conger*), sea bass (*Dicentrarchus labrax*), and octopus (*Octopus vulgaris*), in addition to the shellfishing of stalked barnacle (*Pollicipes pollicipes*). Eel (*Anguilla anguilla*) fishing, and other fishing activities, such as stalked barnacle (*P. pollicipes*) harvesting, play a fundamental role as alternative fishing activities during certain periods when it is difficult and unprofitable to carry out other activities, that is, the first months of the year when many fishing activities cannot be carried out due to bad weather conditions [30].

Artisanal fisheries have traditionally received less research effort than industrial fisheries and have generally received little attention in Europe [31, 32]. This lack of information has reduced the potential for developing effective and integrated management measures aimed at improving the long-term sustainability of artisanal fisheries, considering the complex interactions and linkages between the human and natural dimensions within these fisheries [33]. Despite the comparatively low volume of catches and its economic relevance, artisanal fisheries are important in terms of employment and must be considered in economic terms at the local level. It contributes to strengthening people's attachment to their territory, increasing social stability in rural and peripheral areas [32].

2.2 Traceability as a cornerstone of sustainable fisheries management in Asturias: a case study on scallops (family Pectinidae)

The UN Code of Conduct for Responsible Fisheries stated in 1995 that the "best available scientific data for assessing the state of fishery resources" should be used for effective fisheries management measures [13]. Traditionally, fisheries conservation and management have been conducted on the basis of abundance data, productivity estimates, and information on stock dynamics—that is, an intraspecific group of randomly mating individuals with temporal and spatial integrity [34–37]. However, managers need to be aware that the implementation of legislation, stocking strategies, and other management activities affect the genetic composition of populations [38–40]. Genetic factors play a role in the conservation of fishery resources because fishery resources are the product of their genes, the environment, and the interactions between them [41]. Although understanding of the state of global fisheries has now improved over the last decade, there is a consensus that data remain incomplete, with most of the world's fish stocks lacking formal statistical assessments [16]. In addition, this lack of biological data is compounded by the fact that not all fishery catches are properly reported or recorded by governmental or non-governmental agencies. These unreported catches may be illegal, of unregulated species, or simply not monitored due to logistical barriers [42]. Mislabeling, inaccurate species identification in landings [43, 44], or the modification of the catch area are other factors that also contribute to the unreported exploitation of stocks and the consequent reduction of fishery resources.

The Common Market Organization (CMO) for fishery and aquaculture products is laid down in Regulation (EU) No 1379/2013 of the European Parliament. In Article 35, the commercial designation of the species and its scientific name (among other relevant data) are included as mandatory information on the relevant labeling. Furthermore, Article 37 stipulates that the Member States shall draw up and publish a list of the commercial designations accepted in their respective territories, together with their scientific names [45]. Informative labeling is particularly important for processed products, as any recognizable external morphological characteristics are often removed, so consumers rely on product labeling for information about the contents of the product.

In recent years, molecular biology techniques—based on DNA and sequencing—have gained notoriety in the study of mislabeling and food fraud, allowing species identification even if the product under suspicion is highly processed [46, 47]. A combined assessment of the levels of fraud in the commercialization of fresh and processed specimens of the family Pectinidae, both in retail establishments and restaurants, has been carried out using genetic methods based on mtDNA—16S rRNA gene–and taxonomic methods in Asturias [48]. That research showed that out of 148 samples of 15 commercial products analyzed, 73 samples (49%) and 9 (60%) of the 15 commercial products studied were mislabeled. In the case of the 20 samples purchased in 20 restaurants, all specimens labeled with the common name "zamburiña" were assigned to the Pacific scallop (*A. purpuratus*), resulting in 100% substitution fraud in the samples analyzed. These results are in agreement with the numerous works previously carried out that used molecular markers as an effective tool for the identification of species—both in fresh and frozen products as well as in highly processed ones—that, in many cases, were impossible to identify without the use of this type of tools [49–52]. At the same time as Parrondo et al. study [48] was published, Klapper and Schröder report the development of a multiplex qPCR assay using a TaqMan probe that allowed the rapid and reliable identification of three commercially important scallop species in German supermarkets and fish-mongers—the scallop (*P. maximus*), the Atlantic scallop (*Placopecten magellanicus*), and the Japanese scallop (*Patinopecten yessoensis*) [53]. Using this new tool, Klapper and Schroeder revealed a mislabeling rate of 48%—especially high for products purchased in fish shops. Furthermore, they showed that only 18 out of 33 (52%) samples were correctly labeled and in 12 (36%) samples the scientific name was not provided. Where appropriate, the rate of mislabeling in supermarkets was relatively low (5 out of 21, 24%) compared to fishmongers (8 out of 10, 80%) and restaurants (2 out of 2, 100%) [53]. Despite differences in methodology and target species, the results obtained for Germany by Klapper and Schröder are consistent with the data produced by Parrondo et al. [48] for Asturias.

These investigations join those previously presented by other authors, both in the development or use of forensic biology techniques and in the creation of a threshold of knowledge in the study of mislabeling and fraud of fishery products from marine invertebrates, specifically scallops [54–59]. Despite the difficulty in making accurate estimates of the extent of mislabeling, especially in invertebrates due to lack of data and bias toward certain taxa and geographical areas [60], the high percentages of errors in product labeling found in Parrondo et al. [48] are not unusual in other European countries, such as Iceland, Finland, or Germany, where similar results of between 40 and 50% were obtained [51]. These percentages are alarming if we take into account that most of the samplings carried out for these studies (including the present one) lack temporality, being performed only once in a specific locality, so the approximations tend to be always conservative [49].

Research regarding mislabeling and food fraud in seafood products are increasingly extensive at all levels of the production chain [49, 50, 61], the urgent need for control measures throughout marketing to avoid consumer confusion, mislabeling, or potential health problems—such as allergies due to substitutions [62]—is evident. It is worth mentioning that, in general, there is an increased awareness of the industry to improve the transparency of the food chain, as well as the growth in the number of inspections put in place by European official control bodies, which have achieved a significant reduction in the number of incidences of misdescriptions [63]. This lack of monitoring plays an important role in threatening the sustainability of fisheries, despite international efforts, and may even imply the eventual extinction of more vulnerable overfished species [64].

In Asturias, the existing regulation gives legal and regulatory support to food quality—especially to differentiated quality and organic production—and establishes regulation of infractions and sanctions, with the aim of tackling intrusion and fraud [65]. However, the Asturian law is framed within Community regulations and applies European legislation on labeling. That is why, from a legislative point of view, Article 35 of Regulation (EU) No 1379/2013 of the European Parliament should be amended to include the fishery and aquaculture products listed in points h) and i) of Annex I to that Regulation, which refers to "prepared or preserved fish; caviar and caviar substitutes prepared from fish eggs" and "crustaceans, molluscs, and other aquatic invertebrates, prepared or preserved." It is indisputable that to implement correct management and planning of the exploitation of marine resources and to watch over the rights of consumers, it is necessary to increase routine controls and sanctions—both on fishery products and those prepared and processed throughout the whole production chain—with emphasis on those stages where there is greater evidence of fraud and on those species to which, perhaps, less attention has been paid, such as marine invertebrates.

2.3 Sustainable fisheries management, certifications, and the scarcity of biological data in Asturias

2.3.1 The MSC octopus fishery in western Asturias: dilemma and challenges

Labeling to provide additional ecological information about a product is usually voluntary. FAO recognized that it could contribute to improved fisheries management and convened a technical consultation in 1998, which led to the development of the "Guidelines for the Ecolabeling of Fish and Fishery Products from Marine Capture Fisheries" [66]. Since then, numerous programs have been proposed for ecolabelling seafood products in an effort to encourage fisheries managers to create sustainable fisheries. One of the most recognized today is the Marine Stewardship Council (MSC)—created in 1997, thanks to a collaboration between the World Wildlife Fund (WWF) and Unilever, a multinational company that markets several international brands [67]. The aim of these initiatives is to provide a market-based incentive for sustainable fisheries management. Processors, wholesalers, and retailers who purchase products from these accredited fisheries may acquire the right to affix an eco-label, informing consumers that the product has been caught in a sustainable fishery. Hypothetically, if there were a demand for environmental quality, consumers would respond by purchasing those products with an eco-label, thereby reducing demand for those without and causing price devaluation on unlabeled products. This may result in fishermen putting pressure on fishery managers to achieve sustainability accreditation and thus receive a higher percentage of the price [67].

In February 2016, the MSC awarded the Tapia, El Porto, Ortigueira, and Veiga fishermen's associations the first certification for the octopus fishery worldwide. *O. vulgaris* (Cuvier, 1797) is the cephalopod species with the largest landings in Asturias, being this resource of great importance for the artisanal fleet of the region [68]. This fishery operates with a "management plan for the common octopus

(*O. vulgaris*) in the Principality of Asturias," developed by the Directorate General of Maritime Fisheries of the Government of Asturias with the collaboration of the fishermen's guilds, which is revised every year [69]. From that moment on, all octopus caught by certified vessels is eligible to display the blue MSC ecolabel, allowing consumers to enjoy this species with the security that it comes from a sustainable and environmentally friendly fishery [70]. The MSC report accrediting such sustainability highlighted some weaknesses in certifying this fishery, warning that "biological information on the resource was still scarce" and explicitly recommending that "information on the knowledge of octopus stocks should be improved." This means that relevant parameters, such as maximum sustainable yield and stock definitions, were still lacking or incomplete for octopus fisheries.

Stock assessment is crucial, being an integral part of fisheries management. However, it can be challenging due to the methodological difficulties arising from marine monitoring using traditional methods—such as individual capture (with trawls, nets, or traps) or visual identification of species based on their distinctive morphological characteristics—and the amount of time consumed. Moreover, its financial cost is high, and, in some cases, it is simply unfeasible. The need to overcome these impediments has stimulated the search for new tools and approaches to integrate different environmental dimensions into decision-making in a data-driven policy approach [71].

Since 2016, improvements have been made in those areas where information was lacking for the management of the octopus fishery in Asturias. Particularly important has been the development of a model for octopus stock assessment, and the determination of annual reference points for the fishery, allowing the estimation of an annual TAC per campaign that responds to the situation of the octopus stock in the area [72], as well as making it possible for octopus fished with traps in western Asturias to have achieved—in this year 2021 and for five more years—the recertification of this MSC ecolabel. This work is part of a research project (ECOSIFOOD; MCI-20-PID2019-108481RB-100) funded in 2020 by the Spanish National State Program of Research and Development oriented to the challenges of society.

This last research project is also included as a target to work on obtaining new molecular traceability matrices based on new genomic and environmental DNA (eDNA) data from octopus and their application in temporal and spatial samples to help in detection, quantification, defining historic and contemporary patterns of genetic variation data, stocks and management units. The concept of the biological stock as the basic population unit of exploited species is fundamental to the management of wild fisheries. The delimitation of appropriate conservation units—which is the core of short-term management programs—is also a difficult task in marine systems that have traditionally been characterized as genetically undifferentiated populations [73] due to the large population size, high dispersal potential, and high fecundity of these species [74, 75].

The collection and analysis of water samples for eDNA has, in many cases, proven to be a cost-effective, sensitive, and noninvasive method for species presence/absence surveys, in contrast to established monitoring techniques that rely on the capture of whole organisms [76–78]. Studies now abound listing the many qualities of eDNA analyses for aquatic species detection and distribution assessment using DNA released into the environment by marine organisms, both vertebrates [79–81] and invertebrates [82–84]. The development of more reliable and cost-effective procedures for monitoring commercial species populations may, therefore, improve stock assessment [85]. An eDNA-based method was developed for stock assessment of *O. vulgaris*, pioneering work in the use of this methodology to estimate the sustainability of common octopus fisheries [86]. Furthermore, in that work it was found a positive and significant correlation—Pearson correlation coefficient 0.38627, p-value = 0.01261—between octopus biomass and eDNA abundance detected in tank experiments [86]. These results are in agreement with previous studies that showed a strong linear relationship in aquaria: Takahara et al. found that eDNA concentration was positively correlated with carp biomass in both aquaria and experimental ponds, furthermore, they used this method to estimate carp biomass and distribution in a natural freshwater pond [87]; Maruyama et al. examined the effect of the developmental stage of sunfish (*Lepomis macrochirus*) on environmental DNA release rate, finding a positive relationship between fish size and eDNA release rate [88]; and Klymus et al. used controlled laboratory experiments to measure the amount of eDNA that two invasive carp species (*Hypophthalmichthys nobilis* and *Hypophthalmichthys molitrix*) shed into the water, finding a positive relationship with fish biomass by finding that fed fish, compared to non-fed fish, excrete more eDNA [89].

Other authors have found this relationship to be less pronounced [90–92] or even nonexistent [93] in natural ecosystems. However, in more recent studies, Spear et al. have shown that pikeperch density explains most of the variance in eDNA recovered in lake surface waters in natural systems [94]. Quantification of eDNA abundance is based on the assumption that local population size can be inferred by measuring eDNA concentration at a given locality and that this estimate represents the quantitative relationship between eDNA concentration and underlying population size [95]. However, such a relationship may not always be true, or even present in most cases. The results of Mauvessau et al. [86] show significant variations in the amounts of eDNA detected in the different sampling points located in the Cantabrian Sea. In any case, the observed variation in the amount of eDNA may be due to different and even unknown factors. Compared to eDNA sampling in river systems—which also poses its own set of problems, often difficult to address—factors such as tides, currents, large depths, and rapid movements of individuals in three dimensions may affect the collection of unbiased samples [71, 96].

The results presented by Mauvisseau et al. [86] were obtained using species-specific primers and the qPCR technique using the SYBR Green compound—double-stranded DNA-binding dye that allows the detection of the PCR product as it accumulates during PCR, as it is a simple, easy, and economical option. However, other types of technical approaches are now common, such as qPCR using TaqMan probe or recently digital droplet PCR (ddPCR) [97]. More recently, intraspecific diversity assessments have been performed in several species [98, 99], finding multiple haplotypes that had previously been identified from tissue-derived DNA by Sanger sequencing. This is a revolutionary tool for fisheries and population management, as the use of eDNA could allow detection, quantification, and estimation of diversity with minimal sampling efforts.

The use of environmental DNA-based tools to quantify commercial species populations is of great interest to fisheries managers and policy makers, as stock assessment is a central component of any management and/or conservation program [71]. There is a strong need to inform researchers, advisors, managers, and other stakeholders about the many challenges (and opportunities) associated with the application of environmental DNA analyses in routine marine fisheries management [85].

2.3.2 Getting data from a potentially eco-certifiable fishery in Asturias: the highly valued goose barnacle P. pollicipes

The stalked barnacle fishery in Western Asturias has been co-managed by the fishermen's guilds and the Center for Fishing Experimentation (General Directorate of Maritime Fisheries of the Principality of Asturias) since the eighties. On the west coast of Asturias, harvesting is not open, but rather each fishery guild or group of guilds is granted the right to exclusively exploit barnacles in a strip of the coast, a system known as "exploitation plans." In exchange, the guilds commit to exploiting the barnacle in a sustainable manner (complying with the rules established in the exploitation plan) and to report in minute detail the date, place, and amount of barnacle extracted, which is an extraordinary source of information for research and management [100]. In these systems, fishermen become co-responsible for management, intervening in the design of the exploitation plans. The DGP collects and analyzes the data on daily catches per shellfish gatherer and extraction area collected by the authorized fish rangers in each exploitation plan. At the same time, in collaboration with shellfish harvesters and fish wardens, the total and partial closed seasons and areas for each season are evaluated and proposed.

The potential of genetic approaches for the identification of fish stocks has long been recognized. However, in practice, stock assessments for management purposes often do not incorporate information on the biological stock structure because genetic and biological data are unavailable or ambiguous [101]. Even today, the isolation and characterization of new molecular markers remain difficult and costly for many non-model species [102]. Although microsatellite markers—and now also SNPs—are increasingly available for more species, studies on most marine organisms are still limited by marker availability and biased toward those of greatest commercial interest. A good example of this could be the barnacle fishery, which has an annual economic value of 10 million euros, with about 500 t of landings and 2100 professional fishermen involved [103] and yet has hardly been studied from a genetic point of view [100, 104, 105]. Although this may be alarming due to their importance, crustaceans—and other marine invertebrates as well—still lack genetic and genomic resources compared to other widely studied groups [106, 107]. The research projects PERCEBES (PCIN-2016-120) (funded by the EU Biodiversa call in 2016) and ECOSIFOOD (MCI-20-PID2019-108481RB-100) have targeted this objective of developing new genetic tools (microsatellites and SNPs) to assess genetically the fishery stocks of the stalked barnacle *P. pollicipes*.

Microsatellites are established as the most popular and versatile marker type [17]. Their hypervariable nature confers sufficient power to compare gene pools between populations, as unique alleles appear at low frequencies that are useful for discriminating populations. Under a standard set of parameters that includes 20 highly mutational microsatellite loci and approximately 50 individuals from each of the subpopulations to be sampled, the power to detect deviations from panmixia is very high—even with high gene flow [73]. This is the case of barnacle populations, where patterns of spatial and temporal structuring have been observed at a scale where variability should be homogenized by gene flow through larval dispersal and coined as chaotic genetic patchiness (CGP) [108–110].

The detection of genetic differences between samples far apart in space or time implies the existence of some level of demographic independence and the presence of separate populations [34, 73, 111]. The ongoing analyses carried out with 20 new microsatellite loci aimed to define, with greater precision, the spatiotemporal evolution of the genetic structure of the barnacle *P. pollicipes* [112]. Preliminary results using microsatellites pointed out a population dynamics where *P. pollicipes* displays high genetic diversity along the Iberian Peninsula, which is attributable to large effective population sizes representing a well-connected network of local populations. However, temporal and spatial genetic differentiation of populations over regional scales, on one hand, and a significant reduction in genetic diversity in juveniles, on the other hand, clearly indicate that patterns of exchanges together with seasonal wind-induced upwelling may induce genetic differences between settlers throughout generations. Such patterns of chaotic genetic patchiness are likely due to sweepstake reproductive success with possible collective dispersal or episodic self-recruitment events [112]. In the specific case of the barnacle fishery, the future use of SNP markers would allow a more precise review of concepts such as population structure—which has been described as patchy [112]—and larval connectivity—which on the Asturias coast seems to take place on a small scale [100]; also if there are differences or not between the phenotypes considered of better and worse quality—in previous reports, these differences were not significant [113]—or how landscape components affect the resource on its quality [114]; allowing to know the evolutionary forces that drive the spatio-temporal heterogeneity of barnacles, and thus being able to assess and preserve the evolutionary potential of this fishery resource in a context of global change. Although these boundaries are usually spatially defined, they may also have a temporal component [115].

2.3.3 The management of declining or depleted stocks in Asturias: restock or restore? The sea urchin case study

Paracentrotus lividus is a sea urchin with an important ecological role in the Cantabrian Sea ecosystem. In the same context of global change and triggered in recent decades by overfishing, there has been a definitive decline in sea urchin (*P. lividus*) populations on the Asturian coast. Among the measures to mitigate this population collapse, the Government of Asturias decided to establish a year-round ban on the capture of this species and to undertake a population restoration experience with the aim of preserving this marine resource. However, these practices may entail a series of genetic risks that are widely recognized and documented in the literature and that can be summarized as follows: (i) loss of genetic diversity, (ii) loss of fitness, (iii) change in population composition, and (iv) change in population structure. Moreover, although adverse genetic impacts on wild populations are recognized and documented, little effort is devoted to their actual monitoring [116]. Because of this, these practices are highly controversial [117] and their utility is not always clear to fisheries and conservation goals [118]. Genetic monitoring of *P. lividus* populations in the central Cantabrian Sea (Asturias) was carried out using the mitochondrial DNA cytochrome B gene and microsatellite markers previously described by Calderón et al. [112, 119].

The results of a genetic diversity study based on mtDNA on this marine invertebrate show that the Asturian populations could constitute, at present, a peculiar and so far, undiscovered management unit (MU) of the Bay of Biscay, separate from that of the Atlantic populations. Microsatellite marker data—which reflect recent processes of population dynamics—did not reveal controversial results. However, both markers suggested genetic heterogeneity in the Mediterranean and Candás samples [112]. These results are fundamental, as an essential requirement for sustainable exploitation is the adequacy of biologically relevant processes and the scale of management: mismatches between the two often occur [120]. The *P. lividus* fishery in Galicia has been an example of how the mismatch between biological, fishery, and management scales causes governance failures, leading to overexploitation. *P. lividus* is spatially distributed in nested biological units: patches, micropopulations, local populations, and metapopulations. Fishing operations are local and exploit micro-stocks; however, management units in Galicia often comprise more than one local population. This pattern allowed the depletion of several micro-stocks with no short-term signs in exploitation rates across the managed territory [121]. Identification of MUs is fundamental for the management of natural populations and is crucial for controlling the effects of human activity on species abundance [122]. Local sea urchin populations may be more isolated than suggested by larval dynamics [123–127].

The number of broodstock used in the sea urchin restocking experience carried out so far in Asturias—average number of spawners per event = 14.42—has been clearly far from what is desirable and from the recommended minimum number of broodstock [112]. This causes certain alleles and haplotypes to be overrepresented in the new population, leading to a reduction in the effective population size [128]. Besides this, microsatellite markers used indicated that juvenile individuals used for supplementation were genetically different from wild populations [112]. This means a poor representation of the wild gene pool in broodstock as a result—among other factors—of the low number of individuals used as broodstock since, in the particular case of marine invertebrates, they present a very high fecundity associated with a large variance in reproductive success [129], resulting in small Ne in cultured populations.

In Parrondo et al. [112] work, at least 3.5% overall of the total recaptured sea urchins with hatchery origin were found by randomly sampling 100 juveniles from each of the two restocked localities. Comparison of recapture rates is complicated for pilot studies [118], as they may depend on the objective of the experiment, the number of releases and size of individuals at the time of release, sampling effort, as well as the length of the experimental period and areas surveyed, and also the variance between studies. Despite the already commented previous findings, genetic monitoring of the restocked localities showed that they currently do not differ from the rest of the Asturian localities in terms of genetic diversity using both genetic markers, with no evidence of the Ryman-Laikre effect in the restocked populations [112]. Similar results were found in the *P. maximus* breeding program in Brest Bay (France), which had no effect on genetic diversity [130] and no R-L effect was detected [131]. Even if the reduction in allelic diversity and the alteration of allele frequencies were limited, they could accumulate over generations, gradually eroding the genetic variability of *P. lividus*, so long-term monitoring of these populations is proposed as standard practice.

Habitat restoration—taking into account that kelp forests are in serious decline due to, among other factors, the increase in temperature in the Cantabrian sea that has been occurring since the beginning of the 21st century, the increase in the intensity of storms or the limitation of available nutrients due to changes in the frequency of coastal upwelling [132], the construction of shelters against waves, which also seem to be effective in promoting the colonization of kelp [133], as well as the establishment of marine protected areas (MPAs)—which would allow maintaining the supply of larvae—may be other options to be considered to improve sea urchin populations. In some cases, translocation of adult sea urchins could also be considered; however, translocation can have an impact on the "host" population, so it is necessary to manage the stock of animals to be translocated [133]—many of these measures have been successfully implemented in Japan. The integration of aquaculture-based enhancement with habitat restoration presents a notable opportunity for future research and development [134].

2.3.4 Facing biological invasions and their threat to Asturias exploited fishery resources. The Crepidula case study in the Bay of Biscay

Biological invasions are a key component of the ecological and biodiversity conservation crisis. One of the main threats caused by introduced species is the alteration of the structure of host communities -both terrestrial and aquatic- and the modification of ecosystem functioning [135, 136]. Although only a small fraction of the many species introduced outside their native range are able to thrive and invade new habitats, their impact can be dramatic [9, 137]. The invasion process unfolds as a multistage operation involving the acquisition of a propagule in its native range, the transport of that propagule to the new range, and the introduction, establishment, and spread of the invader in the new habitat [138].

Accurate analysis and effective modeling of current and future distributions of invasive alien species (IAS) are highly dependent on the availability and accessibility of occurrence data and information on the natural history of the species [139]. Because conventional sampling techniques often have very low probabilities of detecting rare species in aquatic ecosystems [140], such as the initial stages of invasion processes, not being really effective until the population is established—years after the first introduction—[141], tools that favor immediacy are necessary to combat the spread and establishment of invasive species, carrying out strategies of "early detection and rapid response" (EDRR).

Early detection is a vital step for the effective management of invasive species. The species-specific molecular markers for *C. fornicata* presented by Miralles et al. can be used to detect the early stages of invasions due to their sensitivity, low cost, and ease and speed of laboratory testing [142]. The results of that research demonstrate the presence of *C. fornicata* in close proximity to the *M. gigas* culture facilities operating in the Eo estuary. This oyster is very robust, with great physiological tolerance and an enormous reproductive potential, qualities that favor its cultivation and that have allowed it to become naturalized in all continents, making its eradication a complex task when it reaches high densities. In addition, it is an engineering species that generates important structural changes in the ecosystems it colonizes. *M. gigas* is also responsible for the global dissemination of many harmful species and pathogens, as is the case of *C. fornicata* [143]. This is why intensive sampling is necessary, as well as management measures to prevent the spread of *C. fornicata*, as *M. gigas* has been previously found in the region [144] even attached to floating marine debris [145], demonstrating its potential disperser of biological invaders [146–148].

2.4 Marine citizenship for the new generations in Asturias? Education as tools for a sustainable fishing strategy

There is a need for an education strategy at all levels for sustainable development that provides knowledge, skills, attitudes, and values to enable citizens to make informed decisions to take responsible action for environmental integrity. Environmental education is necessary to foster behavioral changes in the population that lead to a "citizenship of the sea." People who contribute to this "citizenship of the sea" show awareness and concern for the ocean and are motivated to take personal action to contribute to its protection [149]. The number of youth conservation movements is increasing around the world, for example, Fridays for Future—as the younger generations are more aware of the environmental issues affecting the planet. Consequently, young people seem to play a key role in the development of successful conservation programs. However, other avenues need to be developed to bridge the gap between positive attitudes and a real commitment to conservation and sustainable management [150].

The use of games as tools to enhance the acquisition of technical knowledge has long been studied as a powerful tool for learning, engaging, and tackling complicated tasks [151]. Sustainable sea is a strategy game developed for educational purposes in which players assume the role of fishermen while learning concepts related to the sustainable management of fishery resources [152]. Despite the small sample size, the board game provides information that can be useful for teaching fisheries management. According to the results of the pre- and post-tests, regardless of educational level, all groups improved their knowledge of specific topics after the activity. Knowledge gained by playing an educational game seems to be more effective because it is acquired through hands-on learning [153]. This not only raises awareness of the marine environment and its issues but also encourages a change in values to take personal responsibility for protecting the ocean [154].

The board game can be used to enhance the learning of technical concepts related to marine conservation, fisheries, and sustainable management of marine resources, being an alternative to conventional methods and a more useful educational resource, if possible, in the current context. In 2020, as the COVID-19 pandemic spread across the globe, most countries announced temporary school closures, affecting more than 91% of students worldwide—in April 2020, nearly 1.6 billion children were out of school - [155]. This board game can be very helpful tool when developing Information, Education, and Communication (IEC) activities by teachers. It can be played both in the classroom and among household members, the latter option being interesting because parents/guardians have greater responsibility for making decisions about household practices and this requires greater attention to be paid to how adults and children respond to environmental messages [156]. Marine issues are partly rooted in individual behavioral choices, which, either directly or indirectly through the global marketplace, have the potential to make a significant impact on the marine environment, such as through food choices – choosing correctly labeled or eco-labeled seafood – waste – reducing plastic use – and products [150, 154]. The development of learning can be seen as an intergenerational and multidirectional process that includes (but is not limited to) the information that children bring to families through educational formats for sustainability [156]. In addition, the game can be adapted to other fishing and marine resource scenarios, which are close to the players, as well as different educational levels. In the case of the board game, taking into account its low reproduction cost, it is an affordable tool for schools of all educational levels, as well as for anyone who may be interested in working on the goals of the 2030 Agenda.

3. Take home messages while moving forward to a regional strategy for the management of exploited invertebrates' marine stocks in Asturias

The work summarized in this chapter suggests that advancing toward a sustainable fisheries management that guarantees both, the employability and profitability of the sector, as well as the cultural and natural heritage in the region, is a priority that can only be achieved through R + D + i and educational strategies—which require funding—and the development and implementation of a regional strategy oriented toward sustainable management and exploitation.

3.1 Traceability and control over the products from fishing activities marketed in Asturias are still deficient and must be improved

Irregularities in labeling and high levels of substitution fraud have been found in the analysis of scallops marketed with different degrees of processing and acquired in different establishments. The most processed products are those with the highest proportion of incorrect labeling. There is a need to carry out a review of the legislation and control methodologies—routine inspections, sanctions, etc.—that guarantee consumer rights, as well as the reliability of the databases on the first sale of fishery products on which fishery statistics are based and, therefore, fishery management, so that they become an efficient tool for the establishment of sustainable development strategies in the region.

3.2 Design and establishing of new coastal marine protected areas (MPAs) in Asturias is necessary

MPAs favor the conservation of biodiversity; the protection of critical habitats; the increase of fisheries productivity through the regeneration of populations; the increase of knowledge of the marine environment; the refuge and protection of genetic diversity; and the protection of heritage and cultural diversity [157]. The implementation of Marine Protected Areas (MPAs) in Asturias coasts under the umbrella of Ecosystem-Based Management (EBM) and Integrated Coastal Zone Management (ICZM) has been suggested by different fields, as they are subordinated to the wider ecological, social, economic, and political context of the coastal and oceanic zones of which they form part. The sustainable management and, therefore, the preservation of such relevant natural resources of the region as barnacles, sea urchins, and octopus, would benefit significantly from this. Ecosystem-based management (EBM) and Integrated coastal zone management (ICZM) based on knowledge of fish stocks and the implementation in Asturias of possible marine protected areas (MPA) closer to or including the coast is necessary to ensure an efficient larval supply. On the other hand, it seems advisable to establish the figure of technical personnel to assist in fisheries management in the fishery associations, similar to the existing one in Galicia.

3.3 Start as soon as possible new eco-labeling certification processes for Asturias fisheries

The fishermen's guilds that are part of the co-managed barnacle extraction schemes of central-western Asturias could proceed with an application for MSC eco-sustainability certification. Previous studies endorse the high levels of sustainability of this fishery [103], which are managed through TURFs where fishermen actively participate in all aspects of management and share responsibilities with the administration in decision-making. However, among the disadvantages of this certification are the high costs derived from the external audits necessary to carry out its implementation, as well as the successive recertifications. The standards of this type of certification, which incorporate many aspects that were not considered in traditional management, can be incorporated into the management plans of artisanal fishing resources, being the Administration itself the one that requires them to obtain the "privilege" of exploiting a resource. The application of these sustainability standards to management has been very positive in the octopus experience, so they could be adapted to other well-controlled fisheries, such as stalked barnacles or the extraction of *Gelidium* sp.

3.4 Prevent ecosystem damages due to restocking strategies and think about ecosystem restoration

Mitigation and restoration strategies for the decline of exploited marine populations require genetic control and monitoring programs to confirm that hatchery individual truly represent the wild gene pool and for early detection of possible adverse effects on genetic diversity. Moreover, the mitigation of population decline with autochthonous individuals—as it is being done—is the only possible option, because the use of allochthonous individuals—even those coming from any other population of Atlantic origin—could negatively affect the genetic diversity of wild populations since the new variants could displace the autochthonous ones and affect the adaptability and fitness of local populations. It has become evident that it is extremely important to increase the number of broodstock used to obtain these juveniles. In addition, it is recommended to evaluate the implementation of a habitat restoration plan for sea urchins in Asturias, since this type of combined strategy has proven to be more effective in the recovery of populations.

3.5 Generalize the use and application of genetic tools in the management strategies of the Asturias fisheries

Genetics offers a diverse collection of versatile and useful tools to inform fisheries management on biologically based issues. However, the application of genetic tools is still incipient in many species of fishery interest. Genetic data need to be integrated into the management of fishery resources in Asturias, as they can address issues of direct relevance to the management of these resources; therefore, the implementation of routine genetic studies in management plans is recommended, always from a collaborative approach with managers and taking full advantage of new genetic technologies. The genomic era and the use of eDNA are still waiting to be effectively implemented in the management of marine invertebrates in Asturias.

3.6 Educating new generations in "sustainability" will be a keystone in the Asturias 2030 fisheries development strategy

There is an urgent need to bridge the gap between positive attitudes and real engagement of children and youth in ocean conservation, helping to foster real "citizenship of the sea." Gamification can be an efficient alternative learning method that establishes new knowledge, attitudes, and commitment of the new generations with sustainability in the exploitation of marine resources in Asturias.

Acknowledgements

This work is based on the PhD dissertation *"Genetic tools for the implementation of sustainable management plans in fisheries"* defended by M. Parrondo on December

3, 2021 at the University of Oviedo (Spain). This work was funded by the projects GRUPIN-IDI_2021_000040 and ECOSIFOOD (MCI-20-PID2019-108481RB-I00/AEI/10.13039/501100011033). This is a contribution of the Marine Observatory of Asturias (OMA) and the Biotechnology Institute of Asturias (IUBA).

Author details

Marina Parrondo Lombardía[1], Lucía García-Florez[2], Eduardo Dopico Rodríguez[3] and Yaisel Juan Borrell Pichs[1*]

1 Department of Functional Biology, Genetics, University of Oviedo, Oviedo, Spain

2 Center for Fishing Experimentation, General Directorate of Maritime Fisheries of the Principality of Asturias, Gijón, Spain

3 Department of Education Sciences, University of Oviedo, Oviedo, Spain

*Address all correspondence to: borrellyaisel@uniovi.es

References

[1] Elliot JA. An Introduction to Sustainable Development. 4th ed. London: Routledge; 2012. DOI: 10.4324/9780203844175

[2] ONU-CMMAD. Nuestro Futuro Común (Informe Brundtland). UN; 1987

[3] FAO. The State of World Fisheries and Aquaculture. Contributing to Food Security and Nutrition for all. Rome: FAO Rome; 2016. p. 2016

[4] ONU. La Asamblea General adopta la Agenda 2030 para el Desarrollo Sostenible. Desarro Sosten 2015. Available form: https://www.un.org/sustainabledevelopment/es/2015/09/la-asamblea-general-adopta-la-agenda-2030-para-el-desarrollo-sostenible/ [Accessed: July 17, 2021]

[5] Derraik JGB. The pollution of the marine environment by plastic debris: A review. Marine Pollution Bulletin. 2002;**44**:842-852. DOI: 10.1016/S0025-326X(02)00220-5

[6] Doney SC, Fabry VJ, Feely RA, Kleypas JA. Ocean acidification: The other CO2 problem. Annual Review of Marine Science. 2009;**1**:169-192. DOI: 10.1146/annurev.marine.010908.163834

[7] Jackson JBC. Historical overfishing and the recent collapse of coastal ecosystems. Science. 2001;**293**:629-637. DOI: 10.1126/science.1059199

[8] McCauley DJ, Pinsky ML, Palumbi SR, Estes JA, Joyce FH, Warner RR. Marine defaunation: Animal loss in the global ocean. Science. 2015;**347**:1255641. DOI: 10.1126/science.1255641

[9] Molnar JL, Gamboa RL, Revenga C, Spalding MD. Assessing the global threat of invasive species to marine biodiversity. Frontiers in Ecology and the Environment. 2008;**6**:485-492. DOI: 10.1890/070064

[10] Smith VH, Tilman GD, Nekola JC. Eutrophication: Impacts of excess nutrient inputs on freshwater, marine, and terrestrial ecosystems. Environmental Pollution. 1999;**100**:179-196. DOI: 10.1016/S0269-7491(99)00091-3

[11] FAO. The State of World Fisheries and Aquaculture 2020. Sustainability in Action. Rome: FAO Rome; 2020. DOI: 10.4060/ca9229en

[12] Hilborn R, Amoroso RO, Anderson CM, Baum JK, Branch TA, Costello C, et al. Effective fisheries management instrumental in improving fish stock status. Proceedings of the National Academy of Sciences. 2020;**117**:2218-2224. DOI: 10.1073/pnas.1909726116

[13] FAO. Código de Conducta Para la Pesca Responsable. Rome, Italy: FAO; 1995

[14] Costello C, Ovando D, Hilborn R, Gaines SD, Deschenes O, Lester SE. Status and solutions for the World's unassessed fisheries. Science. 2012;**338**:517-520. DOI: 10.1126/science.1223389

[15] Worm B, Branch TA. The future of fish. Trends in Ecology & Evolution. 2012;**27**:594-599. DOI: 10.1016/j.tree.2012.07.005

[16] Ovando D, Hilborn R, Monnahan C, Rudd M, Sharma R, Thorson JT, et al. Improving estimates of the state of global fisheries depends on better data.

Fish and Fisheries. 2021;**22**(6):1377-1391. DOI: 10.1111/faf.12593

[17] Selkoe KA, Toonen RJ. Microsatellites for ecologists: A practical guide to using and evaluating microsatellite markers. Ecology Letters. 2006;**9**:615-629. DOI: 10.1111/j.1461-0248.2006.00889.x

[18] Sunnucks P. Efficient genetic markers for population biology. Trends in Ecology & Evolution. 2000;**15**:199-203. DOI: 10.1016/S0169-5347(00)01825-5

[19] Thorpe JP, Solé-Cava AM, Watts PC. Exploited marine invertebrates: Genetics and fisheries. Hydrobiologia. 2000;**420**:165-184. DOI: 10.1023/A:1003987117508

[20] Domínguez-Cuesta MJ, Valenzuela P, Rodríguez-Rodríguez L, Ballesteros D, Jiménez-Sánchez M, Piñuela L, et al. Cliff coast of Asturias. In: Morales JA, editor. The Spanish Coastal Systems: Dynamic Processes, Sediments and Management. Cham: Springer International Publishing; 2019. pp. 49-77. DOI: 10.1007/978-3-319-93169-2_3

[21] García de la Fuente L. The artisanal fishing fleet in Asturias: Characterization and quantitative analysis of its economic and social importance. La flota pesquera artesanal en Asturias: Caracterización y análisis cuantitativo de su importancia económica y social. [Thesis Dissertation]. Spain: Universidad de Oviedo; 2020

[22] Álvarez-Fernández E. Humans and marine resource interaction reappraised: Archaeofauna remains during the late Pleistocene and Holocene in Cantabrian Spain. Journal of Anthropological Archaeology. 2011;**30**:327-343. DOI: 10.1016/j.jaa.2011.05.005

[23] Álvarez-Fernández E, Cubas M, Aparicio MT, Cueto M, Elorza M, Fernández P, et al. New data for the late upper Palaeolithic in the Cantabrian region: Arangas cave (Cabrales, Asturias, Spain). Journal of Archaeological Science: Reports. 2020;**29**:102092. DOI: 10.1016/j.jasrep.2019.102092

[24] Álvarez-Fernández E, Bécares J, Jordá Pardo JF, Agirre-Uribesalgo A, Álvarez-Alonso D, Aparicio MT, et al. Palaeoenvironmental and chronological context of human occupations at El Cierro cave (northern Spain) during the transition from the late upper Pleistocene to the early Holocene. Journal of Archaeological Science: Reports. 2020;**29**:102138. DOI: 10.1016/j.jasrep.2019.102138

[25] Álvarez-Fernández E, Ontañón-Peredo R, Molares-Vila J. Archaeological data on the exploitation of the goose barnacle Pollicipes pollicipes (Gmelin, 1790) in Europe. Journal of Archaeological Science. 2010;**37**:402-408. DOI: 10.1016/j.jas.2009.10.003

[26] Álvarez-Fernández E, Andrés M, Aparicio MT, Elorza M, Gabriel S, García-Ibaibarriaga N, et al. Biotic resources in the lower Magdalenian at Cova Rosa (Sardeu, Asturias, Cantabrian Spain). Quaternary International. 2019;**506**:25-34. DOI: 10.1016/j.quaint.2018.06.017

[27] Turrero P, Hórreo JL, López B, Pola IG, Arbizu M, García-Vázquez E. Chronological changes in upper Paleolithic fisheries revealed by museum archival material. PALAIOS. 2013;**28**:228-232. DOI: 10.2110/palo.2012.p12-114r

[28] Ministerio de Agricultura, Pesca y Alimentación. Plan Estratégico Nacional del Fondo Europeo de la Pesca, 2007-2013. CPAGE. Available from: https://cpage.mpr.gob.es/producto/plan-estrategico-nacional-del-fondo-europeo-

de-la-pesca-2007-2013/ (Accessed: May 24, 2021]

[29] García-de-la-Fuente L, González-Álvarez J, García-Flórez L, Fernández-Rueda P, Alcázar-Álvarez J. Relevance of socioeconomic information for the sustainable management of artisanal fisheries in South Europe. A characterization study of the Asturian artisanal fleet (northern Spain). Ocean and Coastal Management. 2013;**86**:61-71. DOI: 10.1016/j.ocecoaman.2013.05.007

[30] González-Álvarez J, García-de-la-Fuente L, García-Flórez L, Del Pino F-RM, Alcázar-Álvarez JL, González-Álvarez J, et al. Identification and characterization of Métiers in multi-species artisanal fisheries. A case study in Northwest Spain. Natural Resource. 2016;7:295-314. DOI: 10.4236/nr.2016.76026

[31] Garcia SM, Allison EH, Andrew NJ, Béné C, Bianchi G, de Graaf G, et al. Towards Integrated Assessment and Advice in Small-Scale Fisheries: Principles and Processes. FAO Fisheries and Aquaculture Technical Paper. No. 515. Rome, FAO. 2008. 84p

[32] Guyader O, Berthou P, Koutsikopoulos C, Alban F, Demanèche S, Gaspar MB, et al. Small scale fisheries in Europe: A comparative analysis based on a selection of case studies. Fisheries Research. 2013;**140**:1-13. DOI: 10.1016/j.fishres.2012.11.008

[33] Bowen RE, Riley C. Socio-economic indicators and integrated coastal management. Ocean and Coastal Management. 2003;**46**:299-312. DOI: 10.1016/S0964-5691(03)00008-5

[34] Carvalho GR, Hauser L. Molecular genetics and the stock concept in fisheries. In: Carvalho GR, Pitcher TJ, editors. Molecular Genetics in Fisheries. Dordrecht: Springer Netherlands; 1995. pp. 55-79. DOI: 10.1007/978-94-011-1218-5_3

[35] Ihssen PE, Booke HE, Casselman JM, McGlade JM, Payne NR, Utter FM. Stock identification: Materials and methods. Canadian Journal of Fisheries and Aquatic Sciences. 1981;**38**:1838-1855. DOI: 10.1139/f81-230

[36] Botsford LW, Castilla JC, Peterson CH. The Management of Fisheries and Marine Ecosystems. Science. 1997;**277**:509-515. DOI: 10.1126/science.277.5325.509

[37] Hilborn R, Orensanz JM, Parma AM. Institutions, incentives and the future of fisheries. Philosophical Transactions of the Royal Society B: Biological Sciences. 2005;**360**(1453):47-57. DOI: 10.1098/rstb.2004.1569

[38] Lorenzen K. Understanding and managing enhancement fisheries systems. Reviews in Fisheries Science. 2008;**16**:10-23. DOI: 10.1080/10641260701790291

[39] Ryman N, Laikre L. Effects of supportive breeding on the genetically effective population size. Conservation Biology. 1991;**5**:325-329. DOI: 10.1111/j.1523-1739.1991.tb00144.x

[40] Waples RS, Punt AE, Cope JM. Integrating genetic data into management of marine resources: How can we do it better? Fish and Fisheries. 2008;**9**:423-449. DOI: 10.1111/j.1467-2979.2008.00303.x

[41] Allendorf F, Ryman N, Utter F. Genetics and fishery management: Past, present, and future. In: Ryman N, Utter F, editors. Population Genetics and Fishery Management. Washington Sea Grant Publications/ University of Washington Press, Seattle and London. Reprinted 2009 by The Blackburn Press, Caldwell, NJ. 1987. pp. 1-19

[42] Cisneros-Montemayor AM, Cisneros-Mata MA, Harper S, Pauly D. Extent and implications of IUU catch in Mexico's marine fisheries. Marine Policy. 2013;**39**:283-288. DOI: 10.1016/j.marpol.2012.12.003

[43] Garcia-Vazquez E, Machado-Schiaffino G, Campo D, Juanes F. Species misidentification in mixed hake fisheries may lead to overexploitation and population bottlenecks. Fisheries Research. 2012;**114**:52-55. DOI: 10.1016/j.fishres.2011.05.012

[44] Vella A, Vella N, Schembri S. A molecular approach towards taxonomic identification of elasmobranch species from Maltese fisheries landings. Marine Genomics. 2017;**36**:17-23. DOI: 10.1016/j.margen.2017.08.008

[45] Reglamento (UE) no 1379/2013. REGULATION (EU) No 1379/2013 OF THE EUROPEAN PARLIAMENT AND OF THE COUNCIL of 11 December 2013 on THE common organization of the markets in fishery and aquaculture products, amending Council regulations (EC) No 1184/2006 and (EC) No 1224/2009 and repealin. Official Journal of the European Union. 2013;**L354**:1-21

[46] Lo YT, Shaw PC. DNA-based techniques for authentication of processed food and food supplements. Food Chemistry. 2018;**240**:767-774. DOI: 10.1016/j.foodchem.2017.08.022

[47] Woolfe M, Primrose S. Food forensics: Using DNA technology to combat misdescription and fraud. Trends in Biotechnology. 2004;**22**:222-226. DOI: 10.1016/j.tibtech.2004.03.010

[48] Parrondo M, López S, Aparicio-Valencia A, Fueyo A, Quintanilla-García P, Arias A, et al. Almost never you get what you pay for: Widespread mislabeling of commercial "zamburiñas" in northern Spain. Food Control. 2021;**120**. DOI: 10.1016/j.foodcont.2020.107541

[49] Horreo JL, Fitze PS, Jiménez-Valverde A, Noriega JA, Pelaez ML. Amplification of 16S rDNA reveals important fish mislabeling in Madrid restaurants. Food Control. 2019;**96**:146-150. DOI: 10.1016/j.foodcont.2018.09.020

[50] Muñoz-Colmenero M, Blanco O, Arias V, Martinez JL, Garcia-Vazquez E. DNA authentication of fish products reveals mislabeling associated with seafood processing. Fisheries. 2016;**41**:128-138. DOI: 10.1080/03632415.2015.1132706

[51] Pardo MÁ, Jiménez E, Viðarsson JR, Ólafsson K, Ólafsdóttir G, Daníelsdóttir AK, et al. DNA barcoding revealing mislabeling of seafood in European mass caterings. Food Control. 2018;**92**:7-16. DOI: 10.1016/j.foodcont.2018.04.044

[52] Pardo MÁ, Jiménez E. DNA barcoding revealing seafood mislabeling in food services from Spain. Journal of Food Composition and Analysis. 2020;**91**:103521. DOI: 10.1016/j.jfca.2020.103521

[53] Klapper R, Schröder U. Verification of authenticity: A rapid identification method for commercial scallop species through multiplex real-time PCR. Food Control. 2021;**121**:107574. DOI: 10.1016/j.foodcont.2020.107574

[54] Abbadi M, Marciano S, Tosi F, De Battisti C, Panzarin V, Arcangeli G, et al. Species identification of bivalve molluscs by pyrosequencing. Journal of the Science of Food and Agriculture. 2017;**97**:512-519. DOI: 10.1002/jsfa.7754

[55] Feng Y, Li Q, Kong L, Zheng X. DNA barcoding and phylogenetic analysis of Pectinidae (Mollusca: Bivalvia)

based on mitochondrial COI and 16S rRNA genes. Molecular Biology Reports. 2011;**38**:291-299. DOI: 10.1007/s11033-010-0107-1

[56] López-Piñón MJ, Insua A, Méndez J. Identification of four scallop species using PCR and restriction analysis of the ribosomal DNA internal transcribed spacer region. Marine Biotechnology. 2002;**4**:495-502. DOI: 10.1007/s10126-002-0030-0

[57] Marín A, Fujimoto T, Arai K. The variable 5′ end of the 16S rRNA gene as a novel barcoding tool for scallops (Bivalvia, Pectinidae). Fisheries Science. 2015;**81**:73-81. DOI: 10.1007/s12562-014-0819-6

[58] Marín A, Villegas-Llerena C, Fujimoto T, Arai K. Novel decaplex PCR assay for simultaneous detection of scallop species with species-specific primers targeting highly variable 5′ end of the 16S rRNA gene. Aquaculture Research. 2017;**48**:920-930. DOI: 10.1111/are.12935

[59] Wen J, Zeng L, Chen Z, Zhou Y, Xu Y, Chen D, et al. Species identification of dried scallop adductor muscle (Yao zhu) products sold on the market. Food Control. 2017;**71**:83-87. DOI: 10.1016/j.foodcont.2016.06.023

[60] Luque GM, Donlan CJ. The Characterization of Seafood Mislabeling: A Global meta-Analysis. Biological Conservation. 2019;**236**:556-570. DOI: 10.1016/j.biocon.2019.04.006

[61] Muñoz-Colmenero M, Klett-Mingo M, Díaz E, Blanco O, Martínez JL, Garcia-Vazquez E. Evolution of hake mislabeling niches in commercial markets. Food Control. 2015;**54**:267-274. DOI: 10.1016/j.foodcont.2015.02.006

[62] Triantafyllidis A, Karaiskou N, Perez J, Martinez JL, Roca A, Lopez B, et al. Fish allergy risk derived from ambiguous vernacular fish names: Forensic DNA-based detection in Greek markets. Food Research International. 2010;**43**:2214-2216. DOI: 10.1016/j.foodres.2010.07.035

[63] Mariani S, Griffiths AM, Velasco A, Kappel K, Jérôme M, Perez-Martin RI, et al. Low mislabeling rates indicate marked improvements in European seafood market operations. Frontiers in Ecology and the Environment. 2015;**13**:536-540. DOI: 10.1890/150119

[64] Agnew DJ, Pearce J, Pramod G, Peatman T, Watson R, Beddington JR, et al. Estimating the worldwide extent of illegal fishing. PLoS One. 2009;**4**:e4570. DOI: 10.1371/journal.pone.0004570

[65] Gobierno del Principado de Asturias. BOPA núm. 47 de 08-III-2019. vol. 47, 2019

[66] FAO. Guidelines for the Ecolabelling of Fish and Fishery Products from Marine Capture Fisheries. Revision 1. Rome, Italy: FAO; 2009

[67] Gudmundsson E, Wessells CR. Ecolabeling seafood for sustainable production: Implications for fisheries management. Marine Resource Economics. 2000;**15**:97-113. DOI: 10.1086/mre.15.2.42629294

[68] Fernández-Rueda P, García-Flórez L. Octopus vulgaris (Mollusca: Cephalopoda) fishery management assessment in Asturias (north-West Spain). Fisheries Research. 2007;**83**:351-354. DOI: 10.1016/j.fishres.2006.10.006

[69] González AF, Macho G, de Novoa J, García M, Macarena N, Silva G. Western Asturias Octopus Traps Fishery of Artisanal cofradías. Bureau Veritas

Iberia, editors. Public Certification Report. Madrid. 2015. 217p

[70] Gestión sostenible del pulpo. n.d. Available from: http://www.pulpodeasturias.es/index.html [Accessed: May 25, 2021]

[71] Gilbey J, Carvalho G, Castilho R, Coscia I, Coulson MW, Dahle G, et al. Life in a drop: Sampling environmental DNA for marine fishery management and ecosystem monitoring. Marine Policy. 2021;**124**:104331. DOI: 10.1016/j.marpol.2020.104331

[72] Roa-Ureta RH, Fernández-Rueda MDP, Acuña JL, Rivera A, González-Gil R, García-Flórez L. Estimation of the spawning stock and recruitment relationship of Octopus vulgaris in Asturias (Bay of Biscay) with generalized depletion models: Implications for the applicability of MSY. ICES Journal of Marine Science. 2021;**78**:2256-2270. DOI: 10.1093/icesjms/fsab113

[73] Waples RS, Gaggiotti O. INVITED REVIEW: What is a population? An empirical evaluation of some genetic methods for identifying the number of gene pools and their degree of connectivity. Molecular Ecology. 2006;**15**:1419-1439. DOI: 10.1111/j.1365-294X.2006.02890.x

[74] Nielsen EE, Kenchington E. A new approach to prioritizing marine fish and shellfish populations for conservation. Fish and Fisheries. 2001;**2**:328-343. DOI: 10.1046/j.1467-2960.2001.00055.x

[75] Ward RD, Woodwark M, Skibinski DOF. A comparison of genetic diversity levels in marine, freshwater, and anadromous fishes. Journal of Fish Biology. 1994;**44**:213-232. DOI: 10.1111/j.1095-8649.1994.tb01200.x

[76] Davy CM, Kidd AG, Wilson CC. Development and validation of environmental DNA (eDNA) markers for detection of freshwater turtles. PLoS One. 2015;**10**:1-13. DOI: 10.1371/journal.pone.0130965

[77] Sigsgaard EE, Carl H, Møller PR, Thomsen PF. Monitoring the near-extinct European weather loach in Denmark based on environmental DNA from water samples. Biological Conservation. 2015;**183**:46-52. DOI: 10.1016/j.biocon.2014.11.023

[78] Turner CR, Miller DJ, Coyne KJ, Corush J. Improved methods for capture, extraction, and quantitative assay of environmental DNA from Asian Bigheaded carp (Hypophthalmichthys spp.). PLoS One. 2014;**9**:e114329. DOI: 10.1371/journal.pone.0114329

[79] Foote AD, Thomsen PF, Sveegaard S, Wahlberg M, Kielgast J, Kyhn LA, et al. Investigating the potential use of environmental DNA (eDNA) for genetic monitoring of marine mammals. PLoS One. 2012;**7**:e41781. DOI: 10.1371/journal.pone.0041781

[80] Thomsen PF, Kielgast J, Iversen LL, Møller PR, Rasmussen M, Willerslev E. Detection of a diverse marine fish Fauna using environmental DNA from seawater samples. PLoS One. 2012;**7**(8):e41732. DOI: 10.1371/journal.pone.0041732

[81] Weltz K, Lyle JM, Ovenden J, Morgan JAT, Moreno DA, Semmens JM. Application of environmental DNA to detect an endangered marine skate species in the wild. PLoS One. 2017;**12**:e0178124. DOI: 10.1371/journal.pone.0178124

[82] Ardura A, Zaiko A, Martinez JL, Samulioviene A, Semenova A, Garcia-Vazquez E. eDNA and specific primers for early detection of invasive

species - a case study on the bivalve Rangia cuneata, currently spreading in Europe. Marine Environmental Research. 2015;**112**:48-55. DOI: 10.1016/j.marenvres.2015.09.013

[83] Forsström T, Vasemägi A. Can environmental DNA (eDNA) be used for detection and monitoring of introduced crab species in the Baltic Sea? Marine Pollution Bulletin. 2016;**109**:350-355. DOI: 10.1016/j.marpolbul.2016.05.054

[84] Wada T, Doi H, Togaki D, Kaida R, Nagano M, Katano I, et al. Exploring a legendary giant squid: An environmental DNA approach. Marine Biology. 2020;**167**:160. DOI: 10.1007/s00227-020-03773-z

[85] Hansen BK, Bekkevold D, Clausen LW, Nielsen EE. The sceptical optimist: Challenges and perspectives for the application of environmental DNA in marine fisheries. Fish and Fisheries. 2018;**19**:751-768. DOI: 10.1111/faf.12286

[86] Mauvisseau Q, Parrondo M, Fernández MP, García L, Martínez JL, García-Vázquez E, et al. On the way for detecting and quantifying elusive species in the sea: The Octopus vulgaris case study. Fisheries Research. 2017;**191**:41-48. DOI: 10.1016/j.fishres.2017.02.023

[87] Takahara T, Minamoto T, Yamanaka H, Doi H, Kawabata ZI. Estimation of fish biomass using environmental DNA. PLoS One. 2012;**7**(4):e35868. DOI: 10.1371/journal.pone.0035868

[88] Maruyama A, Nakamura K, Yamanaka H, Kondoh M, Minamoto T. The release rate of environmental DNA from juvenile and adult fish. PLoS One. 2014;**9**:e114639. DOI: 10.1371/journal.pone.0114639

[89] Klymus KE, Richter CA, Chapman DC, Paukert C. Quantification of eDNA shedding rates from invasive bighead carp Hypophthalmichthys nobilis and silver carp Hypophthalmichthys molitrix. Biological Conservation. 2015;**183**:77-84. DOI: 10.1016/j.biocon.2014.11.020

[90] Schmelzle MC, Kinziger AP. Using occupancy modelling to compare environmental DNA to traditional field methods for regional-scale monitoring of an endangered aquatic species. Molecular Ecology Resources. 2016;**16**:895-908. DOI: 10.1111/1755-0998.12501

[91] Thomsen PF, Møller PR, Sigsgaard EE, Knudsen SW, Jørgensen OA, Willerslev E. Environmental DNA from seawater samples correlate with trawl catches of subarctic, Deepwater fishes. PLoS One. 2016;**11**:e0165252. DOI: 10.1371/journal.pone.0165252

[92] Yamamoto S, Minami K, Fukaya K, Takahashi K, Sawada H, Murakami H, et al. Environmental DNA as a "snapshot" of fish distribution: A case study of Japanese jack mackerel in Maizuru Bay, sea of Japan. PLoS One. 2016;**11**:1-18. DOI: 10.1371/journal.pone.0149786

[93] Spear SF, Groves JD, Williams LA, Waits LP. Using environmental DNA methods to improve detectability in a hellbender (Cryptobranchus alleganiensis) monitoring program. Biological Conservation. 2015;**183**:38-45. DOI: 10.1016/j.biocon.2014.11.016

[94] Spear MJ, Embke HS, Krysan PJ, Zanden MJV. Application of eDNA as a tool for assessing fish population abundance. Environmental DNA. 2021;**3**:83-91. DOI: 10.1002/edn3.94

[95] Chambert T, Pilliod DS, Goldberg CS, Doi H, Takahara T. An analytical framework for estimating aquatic species density from environmental DNA. Ecology and

Evolution. 2018;**8**:3468-3477. DOI: 10.1002/ece3.3764

[96] Canals O, Mendibil I, Santos M, Irigoien X, Rodríguez-Ezpeleta N. Vertical stratification of environmental DNA in the open ocean captures ecological patterns and behavior of deep-sea fishes. Limnology and Oceanography Letters. 2021;**6**(6):339-347. DOI: 10.1002/lol2.10213

[97] Doi H, Uchii K, Takahara T, Matsuhashi S, Yamanaka H, Minamoto T. Use of droplet digital PCR for estimation of fish abundance and biomass in environmental DNA surveys. PLoS One. 2015;**10**:e0122763. DOI: 10.1371/journal.pone.0122763

[98] Parsons KM, Everett M, Dahlheim M, Park L. Water, water everywhere: Environmental DNA can unlock population structure in elusive marine species. Royal Society Open Science. 2018;**5**(8):180537. DOI: 10.1098/rsos.180537

[99] Tsuji S, Maruyama A, Miya M, Ushio M, Sato H, Minamoto T, et al. Environmental DNA analysis shows high potential as a tool for estimating intraspecific genetic diversity in a wild fish population. Molecular Ecology Resources. 2020;**20**:1248-1258. DOI: 10.1111/1755-0998.13165

[100] Rivera A, Weidberg N, Pardiñas AF, González-Gil R, García-Flórez L, Acuña JL. Role of upwelling on larval dispersal and productivity of gooseneck barnacle populations in the Cantabrian Sea: Management implications. PLoS One. 2013;**8**:e78482. DOI: 10.1371/journal.pone.0078482

[101] Ward RD. Genetics in fisheries management. Hydrobiologia. 2000;**420**:191-201. DOI: 10.1023/A:1003928327503

[102] Hauser L, Seeb JE. Advances in molecular technology and their impact on fisheries genetics. Fish and Fisheries. 2008;**9**:473-486. DOI: 10.1111/j.1467-2979.2008.00306.x

[103] Aguión A, Ojea E, García-Flórez L, Cruz T, Garmendia JM, Davoult D, et al. Establishing a governance threshold in small-scale fisheries to achieve sustainability. Ambio. 2022;**51**(3):652-665. DOI: 10.1007/s13280-021-01606-x

[104] Campo D, Molares J, Garcia L, Fernandez-Rueda P, Garcia-Gonzalez C, Garcia-Vazquez E. Phylogeography of the European stalked barnacle (Pollicipes pollicipes): Identification of glacial refugia. Marine Biology. 2010;**157**:147-156. DOI: 10.1007/s00227-009-1305-z

[105] Quinteiro J, Rodríguez-Castro J, Rey-Méndez M. Population genetic structure of the stalked barnacle Pollicipes pollicipes (Gmelin, 1789) in the northeastern Atlantic: Influence of coastal currents and mesoscale hydrographic structures. Marine Biology. 2007;**153**:47-60. DOI: 10.1007/s00227-007-0783-0

[106] Benestan L. Population genomics applied to fishery management and conservation. In: Oleksiak MF, Rajora OP, editors. Population Genomics: Marine Organisms. Cham: Springer International Publishing; 2020. pp. 399-421. DOI: 10.1007/13836_2019_66

[107] Rotllant G, Palero F, Mather PB, Bracken-Grissom HD, Santos MB. Preface: Recent advances in crustacean genomics. Hydrobiologia. 2018;**825**:1-4. DOI: 10.1007/s10750-018-3773-y

[108] Eldon B, Riquet F, Yearsley J, Jollivet D, Broquet T. Current hypotheses to explain genetic chaos under the sea. Current Zoology. 2016;**62**:551-566. DOI: 10.1093/cz/zow094

[109] Hedgecock D, Pudovkin AIAI. SWEEPSTAKES REPRODUCTIVE SUCCESS IN HIGHLY FECUND MARINE FISH AND SHELLFISH : A REVIEW AND COMMENTARY Dennis Hedgecock and Alexander I Pudovkin. Bulletin of Marine Science. 2011;**87**:971-1002. DOI: 10.5343/bms.2010.1051

[110] Johnson MS, Black R. Chaotic genetic patchiness in an intertidal limpet, Siphonaria sp. Marine Biology. 1982;**70**:157-164. DOI: 10.1007/BF00397680

[111] Waples RS. Separating the wheat from the chaff: Patterns of genetic differentiation in high gene flow species. The Journal of Heredity. 1998;**89**:438-450. DOI: 10.1093/jhered/89.5.438

[112] Parrondo M. Herramientas genéticas para la implementación de planes de manejo sostenibles en pesquerías. Genetic tools for the implementation of sustainable management plans in fisheries [Thesis Dissertation]. Universidad de Oviedo; 2021

[113] Seoane Miraz D. Genetic analyses in the gooseneck barnacles (Genus "Pollicipes") [PhD Thesis]. 2015

[114] Rivera A, Gelcich S, García-Flórez L, Acuña JL. Incorporating landscape metrics into invertebrate fisheries management: Case study of the gooseneck barnacle in Asturias (N. Spain). ICES Journal of Marine Science. 2016;**73**:1570-1578. DOI: 10.1093/icesjms/fsw029

[115] Ovenden JR, Berry O, Welch DJ, Buckworth RC, Dichmont CM. Ocean's eleven: A critical evaluation of the role of population, evolutionary and molecular genetics in the management of wild fisheries. Fish and Fisheries. 2015;**16**:125-159. DOI: 10.1111/faf.12052

[116] Laikre L, Schwartz MK, Waples RS, Ryman N. Compromising genetic diversity in the wild: Unmonitored large-scale release of plants and animals. Trends in Ecology & Evolution. 2010;**25**:520-529. DOI: 10.1016/j.tree.2010.06.013

[117] Araki H, Schmid C. Is hatchery stocking a help or harm?. Evidence, limitations and future directions in ecological and genetic surveys. Aquaculture. 2010;**308**:S2-S11. DOI: 10.1016/j.aquaculture.2010.05.036

[118] Kitada S. Economic, ecological and genetic impacts of marine stock enhancement and sea ranching: A systematic review. Fish and Fisheries. 2018;**19**:511-532. DOI: 10.1111/faf.12271

[119] Calderón I, Turón X, Pascual M. Isolation of nine nuclear microsatellites in the common Mediterranean Sea urchin, Paracentrotus lividus (Lamarck). Molecular Ecology Resources. 2009;**9**:1145-1147. DOI: 10.1111/j.1755-0998.2009.02585.x

[120] Reiss H, Hoarau G, Dickey-Collas M, Wolff WJ. Genetic population structure of marine fish: Mismatch between biological and fisheries management units. Fish and Fisheries. 2009;**10**:361-395. DOI: 10.1111/j.1467-2979.2008.00324.x

[121] Ouréns R, Naya I, Freire J. Mismatch between biological, exploitation, and governance scales and ineffective management of sea urchin (Paracentrotus lividus) fisheries in Galicia. Marine Policy. 2015;**51**:13-20. DOI: 10.1016/j.marpol.2014.07.015

[122] Palsbøll PJ, Bérubé M, Allendorf FW. Identification of management units using population genetic data. Trends in Ecology &

Evolution. 2007;**22**:11-16. DOI: 10.1016/j.tree.2006.09.003

[123] Calderón I, Giribet G, Turon X. Two markers and one history: Phylogeography of the edible common sea urchin Paracentrotus lividus in the Lusitanian region. Marine Biology. 2008;**154**:137-151. DOI: 10.1007/s00227-008-0908-0

[124] Calderón I, Pita L, Brusciotti S, Palacín C, Turón X. Time and space: Genetic structure of the cohorts of the common sea urchin Paracentrotus lividus in Western Mediterranean. Marine Biology. 2012;**159**:187-197. DOI: 10.1007/s00227-011-1799-z

[125] Duran S, Palacín C, Becerro MA, Turon X, Giribet G. Genetic diversity and population structure of the commercially harvested sea urchin Paracentrotus lividus (Echinodermata, Echinoidea). Molecular Ecology. 2004;**13**:3317-3328. DOI: 10.1111/j.1365-294X.2004.02338.x

[126] Maltagliati F, Di Giuseppe G, Barbieri M, Castelli A, Dini F. Phylogeography and genetic structure of the edible sea urchin Paracentrotus lividus (Echinodermata: Echinoidea) inferred from the mitochondrial cytochrome b gene. Biological Journal of the Linnean Society. 2010;**100**:910-923. DOI: 10.1111/j.1095-8312.2010.01482.x

[127] Penant G, Aurelle D, Feral JP, Chenuil A. Planktonic larvae do not ensure gene flow in the edible sea urchin Paracentrotus lividus. Marine Ecology Progress Series. 2013;**480**:155-170. DOI: 10.3354/meps10194

[128] Waples RS, Hindar K, Hard J. Genetic risks associated with marine aquaculture. U.S. Dept. Commer., NOAA technical memorandum NMFS-NWFSC. 2012;**119**:149

[129] Hedgecock D. Does Variance in Reproductive Success Limit Effective Population Sizes of Marine Organisms? Genetics and Evolution of Aquatic Organisms. 1994:122-134

[130] Morvezen R, Charrier G, Boudry P, Chauvaud L, Breton F, Strand Ø, et al. Genetic structure of a commercially exploited bivalve, the great scallop Pecten maximus, along the European coasts. Conservation Genetics. 2016;**17**:57-67. DOI: 10.1007/s10592-015-0760-y

[131] Morvezen R, Boudry P, Laroche J, Charrier G. Stock enhancement or sea ranching? Insights from monitoring the genetic diversity, relatedness and effective population size in a seeded great scallop population (Pecten maximus). Heredity. 2016;**117**:142-148. DOI: 10.1038/hdy.2016.42

[132] Álvarez-Losada Ó, Arrontes J, Martínez B, Fernández C, Viejo RM. A regime shift in intertidal assemblages triggered by loss of algal canopies: A multidecadal survey. Marine Environmental Research. 2020;**160**:104981. DOI: 10.1016/j.marenvres.2020.104981

[133] Agatsuma Y. Stock enhancement. In: Lawrence JM, editor. Developments in Aquaculture and Fisheries Science. 2013;**38**:213-224. DOI: 10.1016/B978-0-12-396491-5.00015-0

[134] Taylor MD, Chick RC, Lorenzen K, Agnalt A-L, Leber KM, Blankenship HL, et al. Fisheries enhancement and restoration in a changing world. Fisheries Research. 2017;**186**:407-412. DOI: 10.1016/j.fishres.2016.10.004

[135] Gurevitch J, Padilla D. Are invasive species a major cause of extinctions? Trends in Ecology & Evolution. 2004;**19**:470-474. DOI: 10.1016/j.tree.2004.07.005

[136] Sala OE, Chapin FS, Iii AJJ, Berlow E, Bloomfield J, et al. Global

biodiversity scenarios for the year 2100. Science. 2000;**287**:1770-1774. DOI: 10.1126/science.287.5459.1770

[137] Clavero M, García-Berthou E. Invasive species are a leading cause of animal extinctions. Trends in Ecology & Evolution. 2005;**20**:110. DOI: 10.1016/j.tree.2005.01.003

[138] Sakai AK, Allendorf FW, Holt JS, Lodge DM, Molofsky J, With KA, et al. The population biology of invasive species. Annual Review of Ecology and Systematics. 2001;**32**:305-332. DOI: 10.1146/annurev.ecolsys.32.081501.114037

[139] Simpson A, Jarnevich C, Madsen J, Westbrooks R, Fournier C, Mehrhoff L, et al. Invasive species information networks: Collaboration at multiple scales for prevention, early detection, and rapid response to invasive alien species. Biodiversity. 2009;**10**:5-13. DOI: 10.1080/14888386.2009.9712839

[140] Evans NT, Olds BP, Renshaw MA, Turner CR, Li Y, Jerde CL, et al. Quantification of mesocosm fish and amphibian species diversity via environmental DNA metabarcoding. Molecular Ecology Resources. 2016;**16**:29-41. DOI: 10.1111/1755-0998.12433

[141] Williamson M. Biological invasions. London, UK: Chapman and Hall; 1996. 244p

[142] Miralles L, Parrondo M, Hernández de Rojas A, Garcia-Vazquez E, Borrell YJ. Development and validation of eDNA markers for the detection of Crepidula fornicata in environmental samples. Marine Pollution Bulletin. 2019;**146**:827-830. DOI: 10.1016/j.marpolbul.2019.07.050

[143] Blanchard M. Recent expansion of the slipper limpet population (Crepidula fornicata) in the bay of Mont-saint-Michel (Western Channel, France). Aquatic Living Resources. 2009;**22**:11-19. DOI: 10.1051/alr/2009004

[144] Miralles L, Ardura A, Arias A, Borrell YJ, Clusa L, Dopico E, et al. Barcodes of marine invertebrates from north Iberian ports: Native diversity and resistance to biological invasions. Marine Pollution Bulletin. 2016;**112**:183-188. DOI: 10.1016/j.marpolbul.2016.08.022

[145] Miralles L, Gomez-Agenjo M, Rayon-Viña F, Gyraitė G, Garcia-Vazquez E. Alert calling in port areas: Marine litter as possible secondary dispersal vector for hitchhiking invasive species. Journal for Nature Conservation. 2018;**42**:12-18. DOI: 10.1016/j.jnc.2018.01.005

[146] Ibabe A, Rayón Viña F, Martínez J, Garcia-Vazquez E. Environmental DNA from plastic and textile marine litter detects exotic and nuisance species nearby ports. PLoS One. 2020;**15**:e0228811. DOI: 10.1371/journal.pone.0228811

[147] Rech S, Salmina S, Borrell YJ, García-Vázquez E. Dispersal of alien invasive species on anthropogenic litter from European mariculture areas. Marine Pollution Bulletin. 2018;**131**:10-16. DOI: 10.1016/j.marpolbul.2018-03.038

[148] Rech S, Thiel M, Borrell YJ, García-Vázquez E. Travelling light: Fouling biota on macroplastics arriving on beaches of remote Rapa Nui (Easter Island) in the South Pacific Subtropical Gyre. Marine Pollution Bulletin. 2018;**137**:119-128. DOI: 10.1016/j.marpolbul.2018.10.015

[149] Mckinley E. A critical evaluation of the application of marine citizenship in sustainable marine management in the UK. [Thesis Dissertation]. Bournemouth University; 2010

[150] García-Gallego A, Borrell YJ, Miralles L. The future of marine citizenship is now: Cetacean conservation in the eyes of young Spanish citizens. Aquatic Conservation: Marine and Freshwater Ecosystems. 2021;**31**:3146-3155. DOI: 10.1002/aqc.3696

[151] Hoffman B, Nadelson L. Motivational engagement and video gaming: A mixed methods study. Educational Technology Research and Development. 2010;**58**:245-270. DOI: 10.1007/s11423-009-9134-9

[152] Parrondo M, Rayon-Viña F, Borrell YJ, Miralles L. Sustainable Sea: A board game for engaging students in sustainable fisheries management. Applied Environmental Education and Communication. 2021;**20**:406-421. DOI: 10.1080/1533015X.2021.1930608

[153] Sharp LA. Stealth learning: Unexpected learning opportunities through games. Journal of Instructional Research. 2012;**1**:42-48

[154] McKinley E, Fletcher S. Improving marine environmental health through marine citizenship: A call for debate. Marine Policy. 2012;**36**:839-843. DOI: 10.1016/j.marpol.2011.11.001

[155] UNESCO. Education: From disruption to recovery. UNESCO. 2020. Available from: https://en.unesco.org/covid19/educationresponse [Accessed: September 5, 2021]

[156] Walker C. Tomorrow's leaders and Today's agents of change? Children, sustainability education and environmental governance. Children and Society. 2017;**31**:72-83. DOI: 10.1111/chso.12192

[157] Cicin-Sain B, Belfiore S. Linking marine protected areas to integrated coastal and ocean management: A review of theory and practice. Ocean and Coastal Management. 2005;**48**:847-868. DOI: 10.1016/j.ocecoaman.2006.01.001

Chapter 6

Plasma and Hemocyanin Phenoloxidase Derived from the Hemolymph of Giant Freshwater Prawn *Macrobrachium rosenbergii* (De Man, 1879)

Mullaivanam Ramasamy Sivakumar and Rangasamy Shanthi

Abstract

We attempted to study the immune response in *M. rosenbergii* by melanization reaction produced by plasma phenoloxidase (PO) activity. The substrate affinity of the PO enzyme was determined using different phenolic substrates, and it was found that the diphenols were only oxidized. The enzyme was characterized as catechol oxidase type of PO and L-3,4 dihydroxyphenylalanine (L-DOPA) showed the highest substrate affinity to the enzyme. The biochemical parameters that determined optimum enzyme activity were found to be 2.5 mM L-DOPA at an absorbance of 470 nm, 10 mM Tris–HCl buffer at pH 7.5, temperature at 25°C, and 15 min incubation. Kinetic characteristics of plasma were studied from the *M. rosenbergii*. The hemocyanin was isolated by gel filtration chromatographic technique using Sephadex G-100. The *M. rosenbergii* hemocyanin (*Mr*HC) showed only one band with a molecular weight of 325 kDa on native polyacrylamide gel electrophoresis (PAGE) when stained with Coomassie Brilliant Blue (CBB) and bathocuproine sulfonic acid. The reduction of *Mr*HC protein in SDS-PAGE displayed three subunits with a molecular weight of 74, 76, and 78 kDa, respectively. Determination of optimal condition for PO activity of plasma has also been attempted. The plasma optimal condition taken for the *Mr*HC was tested for its ability to oxidize diphenols such as L-DOPA was shown only PO activity. These results showed that in the presence of PO and peroxidase inhibitors, phenylthiourea (PTU) and tropolone respectively have decreased plasma and *Mr*HC PO activity. This indicates that hemocyanin triggers innate immunity probably through one of its subunits that function as the active moiety.

Keywords: innate immunity, phenolic substrates, phenoloxidase, hemocyanin, kinetics, inhibitors

1. Introduction

Aquaculture is the fastest-growing, food-producing, profitable, and one of the major employment generating sectors in coastal areas and is expected to quintuple in the coming 50 years [1]. The giant freshwater prawn, *Macrobrachium rosenbergii* known as scampi, is an important aquaculture species in tropical and subtropical regions with the immense commercial export market value; however, the production of the prawns is curtailed by diseases caused by opportunistic pathogens in the rearing environment [2]. Resistance to diseases is based on the strategic improvement of the immune system of the animal and requires intensive research on immune components and its function [3]. As for any invertebrate, the innate immune mechanism in prawns includes cellular [4, 5] and humoral components [6–9].

Invertebrates lack adaptive immunity; therefore, they completely depend on innate immune systems for host defense. Melanization, which is a major innate defense system in invertebrates, is controlled by the enzyme phenoloxidase (PO) [10–12]. The active PO is a bifunctional enzyme that catalyzes the *o*-hydroxylation of monophenols to catechols and the oxidation of *o*-diphenols into *o*-quinones [13]. The first reaction involves monophenolase activity, which converts tyrosine to L-DOPA, which is then oxidized to quinone by diphenols activity of PO [13, 14]. The resulting quinones are converted to melanin by a series of intermediate steps involving enzymatic and nonenzymatic reactions [15, 16]. In one of these enzymatic reactions, dopachrome is decarboxylated by dopachrome isomerase (also called dopachrome tautomerase or Dopachrome Conversion Factor) to form dihydroxy indole, which is then converted to melanin [17–19].

The immune response in crustaceans mainly depends on nonspecific immunity, involving the cellular immunity of hemocytes [20] and humoral immunity through phenoloxidase [8] and agglutination [9]. In insects and crustaceans, phenoloxidase usually exists as a nonactive zymogen, prophenoloxidase (proPO), whose activation to the PO form is tightly regulated via an enzymatic cascade because the melanization reaction generates toxic compounds such as quinone species. This cascade is triggered by the presence of several microbial cell wall components such as β-1,3-glucan, lipopolysaccharides, and peptidoglycan [10]. There is a detectable or high amount of PO activity in crustacean plasma [8, 21, 22] that could be derived from proPO released from hemocytes [22] or from hemocyanin [8, 23, 24] for melanization activity, which remains unclear.

The present study attempted to characterize plasma PO activity in terms of substrate specificity, optimum ionic strength, pH, temperature, and incubation time to determine the biochemical and physiological conditions that support enzyme activity. Furthermore, to understand the substrate affinity of the plasma PO enzyme activity, the kinetics of the enzyme's rate of reaction was determined in the Lineweaver-Burk plot. There is evidence to show that the kinetics of the crustaceans phenoloxidases vary among the different components of the hemolymph as well as species [25–28]. Hence, an attempt has been made to optimize the conditions for determining PO activity of plasma including Km and Vmax value of freshwater prawn *M. rosenbergii*. Based on the determination of optimal condition, PO activity in *Mr*HC (325 kDa) has also been attempted.

2. Materials and methods

2.1 Experimental freshwater prawns

Adult intermoult of the giant freshwater prawns, *M. rosenbergii* weighing around 25–30 ± 1.58 g were purchased from Aqua Farm, Kalpakkam, Kanchipuram District, Tamil Nadu, India. The prawns were carefully transported to the laboratory maintained in 500 L FRP tanks (10 no of tanks, per tank 25 animals) containing at room temperature (28°C + 1°C) continuously aerated freshwater, which was changed thrice a week. The prawns were fed with egg white *ad libitum* and were acclimatized to the laboratory conditions for at least 4–5 days before use. Twenty-five percent of water was renewed daily to remove the unfed and fecal materials. The uninjured, intermoult animals were used throughout this study.

2.2 Hemolymph collection and preparation of plasma

Hemolymph (100 μl) was collected by cardiac puncture using a 23G needle attached to a clean, sterile plastic syringe containing 1.9 ml of ice cold iso-osmotic buffer, TBS-I (Tris 50 mM, NaCl 210 mM, KCl 5 mM, MgCl2 2.5 mM, pH 7.5) mixed and centrifuged in a pre-chilled polypropylene tube (161 x g, 8 min, 4°C) to obtain 1.5 ml of the supernatant as plasma. The exclusion of hemocytes was verified in the collected plasma by observation under phase-contrast microscope. About 50 prawns (each determination, N = 50) were required for collection of 100 μl acellular plasma, following Sivakumar et al. [8].

2.3 Oxidation of phenolic substrates

We tested the oxidative activity of 0.1 ml plasma was tested by incubating with 1.9 ml of different phenolic substrate solutions (5 mM tyrosine, tyramine, L-DOPA, DL-DOPA, dopamine, catechol, hydroquinone, and pyrogallol) for 20 min at 25°C. The color developed was measured spectrophotometrically (Shimadzu UV-160A spectrophotometer, Kyoto, Japan) at 300–700 nm against a reagent blank in which suitable substrates were substituted for plasma.

2.4 Effect of different concentrations of L-DOPA

To 0.1 ml of plasma was mixed 1.9 ml of L-DOPA at different concentrations (1–10 mM) and incubated for 20 min at 25°C. The color developed was measured spectrophotometrically at 470 nm against a reagent blank (L-DOPA).

2.5 Effect of ionic strength on oxidation of L-DOPA

The effect of buffer ionic strength on oxidation of L–DOPA by plasma was assessed by incubating 0.1 ml plasma with 1.9 ml of 2.5 mM L–DOPA prepared in different ionic strength (5–100 mM) at 25°C. After 20 min, the optical density of each of these reaction mixtures was determined spectrophotometrically at 470 nm against a reagent blank (L-DOPA).

2.6 Effect of pH on oxidation of L-DOPA

The ability of plasma to oxidize L-DOPA at different *p*H was tested by incubating 0.1 ml of plasma with 1.9 ml of a substrate of (2.5 mM L-DOPA) solutions prepared in 10 mM Tris–HCl buffer at different pH (6.0–9.0) for 20 min at 25°C. The color developed was measured spectrophotometrically at 470 nm against a reagent blank (L-DOPA).

2.7 Oxidation of L-DOPA exposed to different temperature

Effect of different temperature was tested by incubating 0.1 ml of plasma with 1.9 ml of substrate (2.5 mM L-DOPA) solutions prepared in 10 mM Tris–HCl (pH 7.5) buffer at a different temperature ranging from 10 to 90°C for 20 min. The color developed was measured spectrophotometrically at 470 nm against a reagent blank (L-DOPA).

2.8 Effect of various time intervals on L-DOPA

To 0.1 ml of plasma was mixed 1.9 ml of 2.5 mM L-DOPA (10 mM Tris–HCl; pH 7.5) and incubated for different time intervals (5–30 min) at 25°C. The color developed was measured spectrophotometrically at 470 nm against a reagent blank (L-DOPA).

2.9 Kinetic parameters, km, and Vmax of plasma phenoloxidase enzyme

To measure the kinetic parameters of plasma PO enzyme, different concentrations of L-DOPA (1.0–10.0 mM) were mixed with 0.1 ml of plasma and incubated for 15 min and absorbance was read at 470 nm. Michaelis–Menten constant was estimated by plotting substrate concentrations [S] and rate of PO activity [V]. Lineweaver-Burk plot was plotted as reciprocal of substrate concentration [1/S] and rate of PO activity [1/V]. The resultant plot is given a line that intercepted X-axis to give −1/Km value and intercepted the Y-axis to give 1/Vmax. The slope Km/Vmax was determined, and the resultant plot was rechecked using Eq. $Y = mx + c$.

2.10 Partial purification of hemocyanin

To 50 ml of plasma was centrifuged and dialyzed (MW exclusion limit <14,000 kDa and > 12,000 kDa) extensively against TBS-II (Tris 10 mM, NaCl 200 mM, $CaCl_2$ 10 mM; pH 7.5). Then the dialyzed plasma was ultracentrifugation at 200,000 xg for 180 min at 4°C (Beckman LE-80; Beckman Coulter, Brea, CA, USA). After ultracentrifugation, the supernatant was decanted and the pellet, which is made of hemocyanin, was collected and dissolved in TBS-II and used freshly for further purification.

2.11 Purification of *Mr*HC

To purify the *Mr*HC, a Sephadex G-100 (Sigma-Aldrich; bead diameter: 40–120 μm) column (36 x 1.6 cm; XK 16, Pharmacia, Uppsala, Sweden) was prepared using gel filtration chromatography technique and thoroughly equilibrated with TBS-II. The hemocyanin collected by ultracentrifugation from the plasma was passed through the Sephadex G-100 column at a flow rate15 $ml.h^{-1}$. The purified fractions were continuously monitored for absorbance at 280 nm and 1 ml fractions were collected. The collected protein samples were stored at −80°C for further analyses.

2.12 Determination of protein

The protein content in the plasma and purified *Mr*HC (325 kDa) samples were determined according to Bradford [29] using bovine serum albumin as the standard. All chemicals used in the study were purchased from Sigma-Aldrich, St. Louis, MO, USA.

2.13 Electrophoretic analysis

The protein profiles of plasma and purified *Mr*Hc were analyzed in discontinuous polyacrylamide gel electrophoresis (PAGE) under non-denaturing conditions [30]. This was performed using 3% stacking gel (pH 6.7) and a 7% separating gel in Tris-glycine buffer (pH 8.9). Electrophoresis was performed at a constant current of 3 mA/sample at 10°C for 2 h on a slab gel measuring 8 x 8 cm. The gels were stained with Coomassie brilliant blue (CBB) R-250 (GE Health Care Biosciences, Tamil Nadu, India) or bathocuproine sulfonic acid following the methods of Maurer [30] and Bruyninckx et al. [31].

The molecular masses of the purified *Mr*Hc were estimated using sodium dodecyl sulfate–polyacrylamide gel electrophoresis (SDS-PAGE) with 5 and 10% polyacrylamide as the stacking and separating gel, respectively, following the method of Laemmli [32]. The purified *Mr*Hc subunits were visualized by staining the gel with CBB R-250. The molecular weight of the purified *Mr*Hc subunits was calculated using molecular marker proteins (GE Health Care Biosciences, Tamil Nadu, India).

2.14 Oxidation of diphenolic substrates by purified *Mr*HC

We tested the oxidative activity of 40 μl purified *Mr*HC by incubating with 160 μl of different phenolic substrate solutions (2.5 mM L-DOPA, DL-DOPA, dopamine, and catechol) for 15 min at 25°C. The color developed was measured spectrophotometrically at 300–700 nm against a reagent blank in which Tris–HCl buffer (10 mM, pH 7.5) was substituted for purified *Mr*HC.

2.15 Phenoloxidase activity

The PO activity of plasma (0.1 ml) or purified *Mr*HC (40 μl) was incubating with 1.9 ml or 160 μl of phenolic substrate solution (2.5 mM L–DOPA) for 15 min at 25°C. After incubation, the color developed was measured spectrophotometrically (Shimadzu UV-160A spectrophotometer, Japan) or using ELISA plate reader (BioTek, PowerWave XS, USA) at an absorbance of 470 nm against a reagent blank of substrate solution (L–DOPA).

2.16 Effect of inhibitors on PO activity

In this experiment, 0.1 ml of plasma or 40 μl of purified *Mr*HC was mixed with an equal volume of inhibitors. 3 mM Phenylthiourea (PTU) or 16 mM tropolone containing 16 mM H_2O_2 was preincubated for 10 min at 25°C. An aliquot of 0.2 ml or 80 μl of these reaction mixtures from control or experiments was incubated with 1.8 ml or 120 μl of 2.5 mM L-DOPA for 15 min at 25°C. After incubation, the color developed was measured spectrophotometrically or ELISA plate reader at an absorbance of 470 nm against a reagent blank of substrate solution.

2.17 Statistical analysis

The data were expressed as mean ± SD of triplicate experiments from five determinations. Statistical analyses were done using SPSS software (version 20; SPSS, New York, USA). The variation between experimental and control was evaluated by one-way analysis of variance (ANOVA) and significance was assessed at 0.01 probability ($^{**}p < 0.01$).

3. Results

3.1 Effect PO activity with various substrates

The plasma separated from the hemolymph of the freshwater prawn *M. rosenbergii* was tested for PO activity with various phenolic substrates. Among the diphenolic substrates, the plasma showed the highest activity with L–DOPA (470 nm) when compared with DL-DOPA (440 nm), dopamine (440 nm), and catechol (470 nm) as shown in **Figure 1**. However, the monophenols including tyramine and L-tyrosine or polyphenols such as hydroquinone and pyrogallol failed to show any oxidation by plasma. Since the highest oxidative activity was obtained with L-DOPA, this substrate was used to detect PO activity in all subsequent experiments performed in this study.

3.2 Effect of substrate concentration on PO activity

The plasma PO activity was tested with different concentrations of L-DOPA (1.0–10.0 mM), and the PO activity was found to be higher with L-DOPA at a

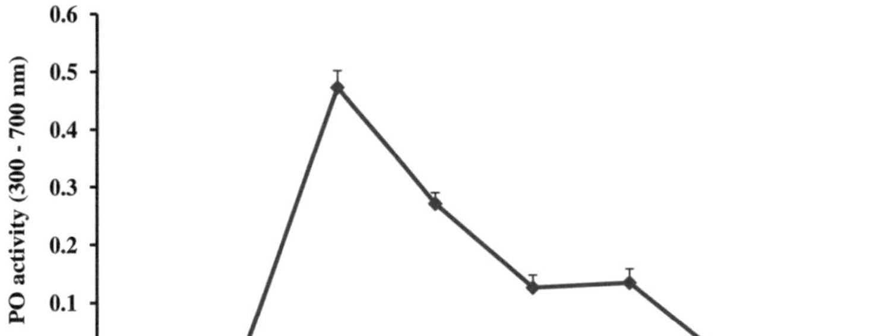

Figure 1.
PO activity of plasma with different phenolic substrates (5 mM) in Tris buffer (Tris–HCl 50 mM, pH 7.5) incubated at 25°C for 20 min and absorbance at 300–700 nm. The PO activity in optical density obtained at absorbancy maxima of respective substrates. Data represents mean of triplicate repeats of five determinations (mean ± SD) in the same way in all further experiments.

concentration of 2.5 mM than that of 1 mM or higher concentrations (5.0, 7.5, and 10.0 mM) as shown in **Figure 2**. This experiment clearly suggested that the optimum concentration for testing PO activity in plasma was 2.5 mM of L-DOPA.

3.3 Effect of ionic strength

The PO activity of plasma was tested with Tris–HCl buffer (pH 7.5) of different ionic strengths (5–100 mM), and the highest PO activity was found with 10 mM Tris–HCl buffer when compared with other ionic strengths tested as shown in **Figure 3**. This result recommended that the optimum concentration for testing PO activity in plasma was 10 mM of Tris–HCl buffer.

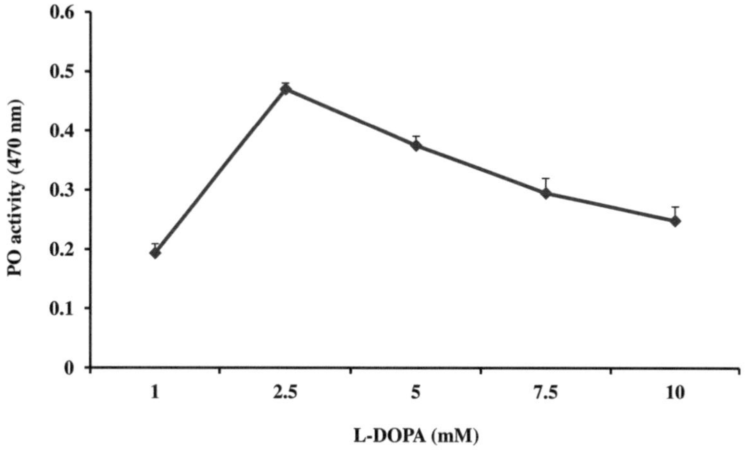

Figure 2.
Effect of different concentrations of substrate (L-DOPA) by plasma phenoloxidase activity of freshwater prawn M. rosenbergii.

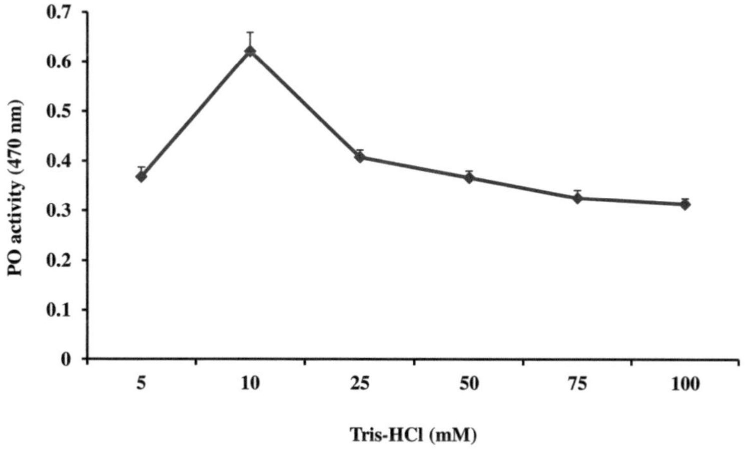

Figure 3.
Effect of ionic strength of Tris–HCl buffer on oxidation of L-DOPA (2.5 mM) by plasma phenoloxidase activity of freshwater prawn M. rosenbergii.

3.4 Effect of optimum pH

The PO activity of plasma was assessed by oxidation of L-DOPA at various pH values ranging from 6.0 to 9.0, pH above 7.5 showed the brown color formation of dopachrome. The PO activity was decreased at pH 6.0–7.0 and 8.0–9.0; thus, pH 7.5 was taken as the optimum pH for the study of plasma PO activity (**Figure 4**).

3.5 Effect of optimum temperature

The PO activity of plasma was demonstrated by performing oxidation of 2.5 mM L-DOPA in the presence of 10 mM Tris–HCl at a pH 7.5. The reaction mixture was incubated for 20 min at different temperatures ranging from 10 to 90°C. The PO activity was stable and attained a peak at 25°C, which was taken as an optimum temperature for PO activity. At temperature below or above 25°C, a decline in PO activity was observed (**Figure 5**).

3.6 Effect of time intervals

The PO activity of plasma was evaluated by performing oxidation of 2.5 mM L-DOPA in the presence of 10 mM Tris–HCl at a pH 7.5 and temperature 25°C at various incubation periods ranging from 5 to 30 min. The maximum PO activity was at 15 min, which was determined as the optimum incubation time (**Figure 6**).

3.7 Kinetic behavior

The kinetic characteristics of plasma PO activity were determined from the rate of the reaction, which was calculated from the oxidation of L-DOPA at different concentrations (1.0–10.0 mM) in 15 min. The Michaelis–Menten constant Km was calculated to be 0.75, and maximum velocity (Vmax) was found to be 0.58 as shown in **Figure 7A**. Application of Km and Vmax yielded Lineweaver-Burk plot with a line slope of 1.2, which on extrapolation intercepted at −1.3 that was plotted as −1/Km and on Y-axis 1/Vmax was derived at 1.7 on X-axis (**Figure 7B**).

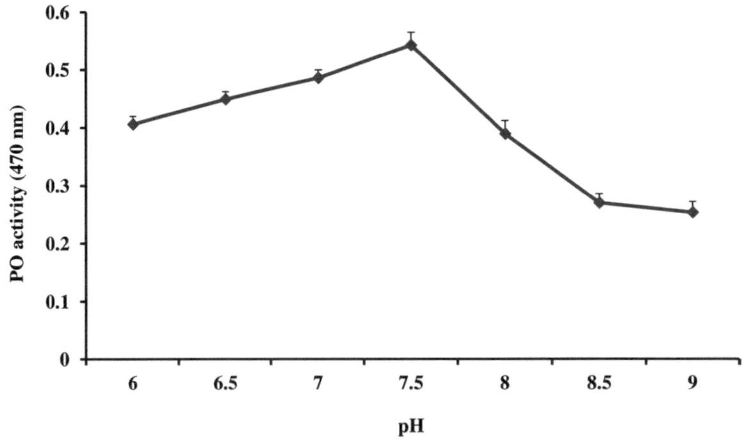

Figure 4.
Effect of pH on oxidation of L-DOPA (2.5 mM), Tris–HCl buffer (10 mM) by plasma phenoloxidase activity of freshwater prawn M. rosenbergii.

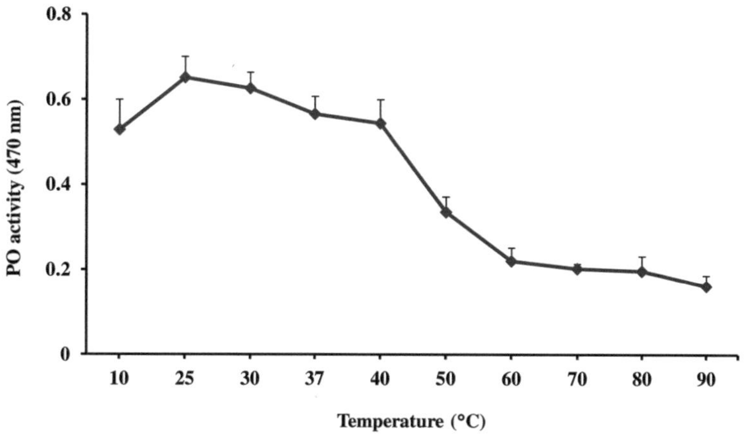

Figure 5.
Effect of temperature on oxidation of L-DOPA (2.5 mM), Tris–HCl buffer (10 mM), pH 7.5 by plasma phenoloxidase activity of freshwater prawn M. rosenbergii.

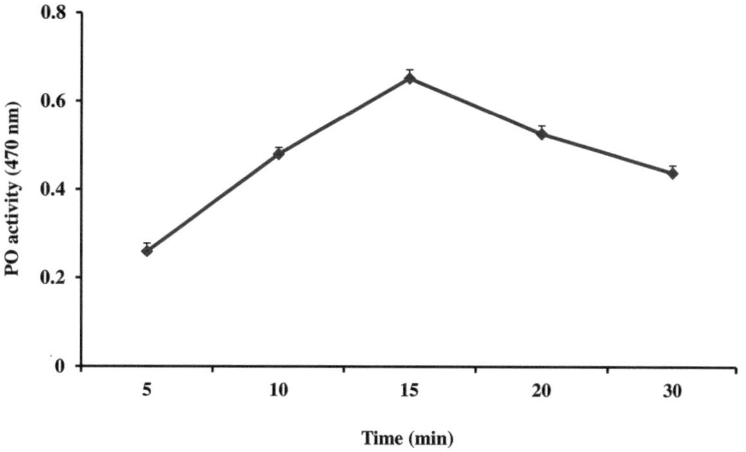

Figure 6.
Effect of incubation time on oxidation of L-DOPA (2.5 mM), Tris–HCl buffer (10 mM, pH 7.5) by plasma phenoloxidase activity of freshwater prawn M. rosenbergii.

3.8 Purification of hemocyanin from the plasma of *M. rosenbergii*

The hemocyanin was loaded on the Sephadex G-100 column for gel filtration chromatographic separation, and the purified *Mr*HC peak fractions were collected at an absorbance of 280 nm (Fiure 8A). Hemocyanin protein was identified on PAGE–7% by CBB staining as distinct single bands of molecular weight 325 kDa (**Figure 8B**; lane 2). Staining with BCSA affirmed that the proteins contained copper and represented the copper containing proteins of hemocyanin (**Figure 8B**; lane 3). The chromatographic separation with electrophoretic observations of the separated proteins clearly indicated the occurrence of hemocyanin in *M. rosenbergii* as single separate copper-containing protein. As shown in **Figure 8C**, the purified *Mr*HC protein after reduction in SDS-PAGE (10%) cleaves into three subunits of 74, 76, and 78 kDa molecular mass, respectively (lane 2).

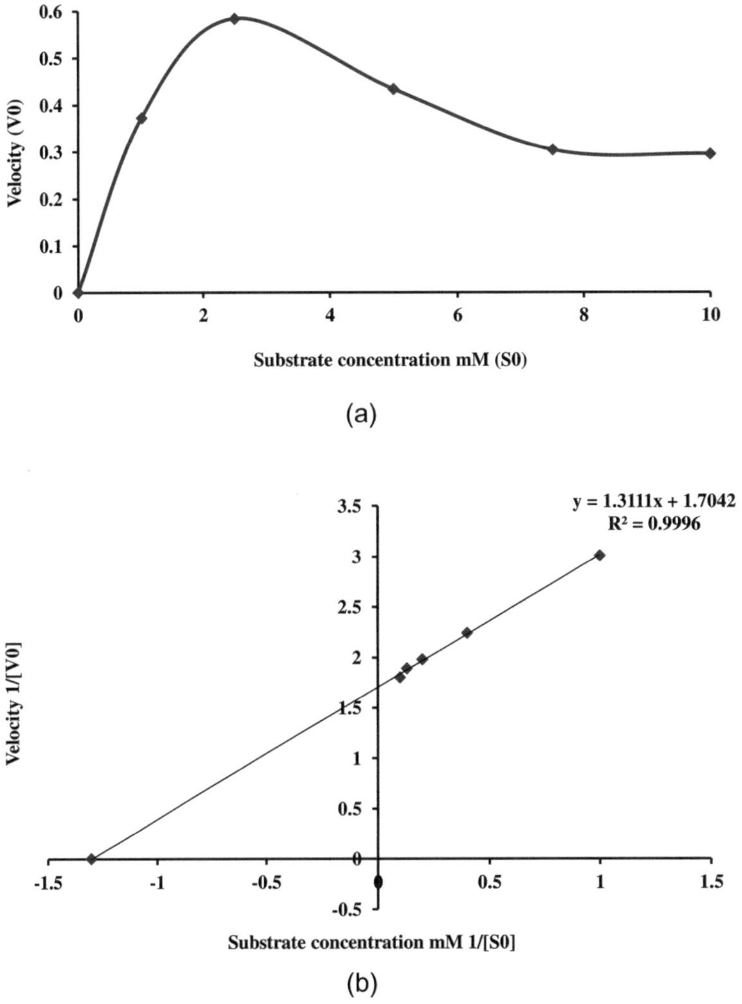

Figure 7.
(A) Kinetic properties of PO activity in plasma of M. rosenbergii at different substrate concentrations of L-DOPA as shown in Michaelis–Menten curve. (B) The K_m *and* V_{max} *values are calculated using Lineweaver-Burk plot of PO activity in plasma of M. rosnbergii with L-DOPA as substrate.*

3.9 Phenoloxidase activity with diphenolic substrates in *Mr*HC

The purified *Mr*HC (325 kDa) was tested for PO activity with diphenolic substrates. Among the substrates, the purified *Mr*HC (325 kDa) showed activity only with L-DOPA while with the other diphenols, such as DL-DOPA, dopamine, and catechol, it failed to show any oxidation activity. Since the highest oxidative activity was detected with L-DOPA (**Figure 9**), this substrate was used for the determination of phenoloxidase activity and inhibition study.

3.10 Effect of PO inhibitors on oxidation of L-DOPA by plasma and purified *Mr*HC

Pretreatment of plasma or purified *Mr*HC with PTU (3 mM) decreased the oxidation of 2.5 mM L-DOPA compared with control, and the reduction was found to be

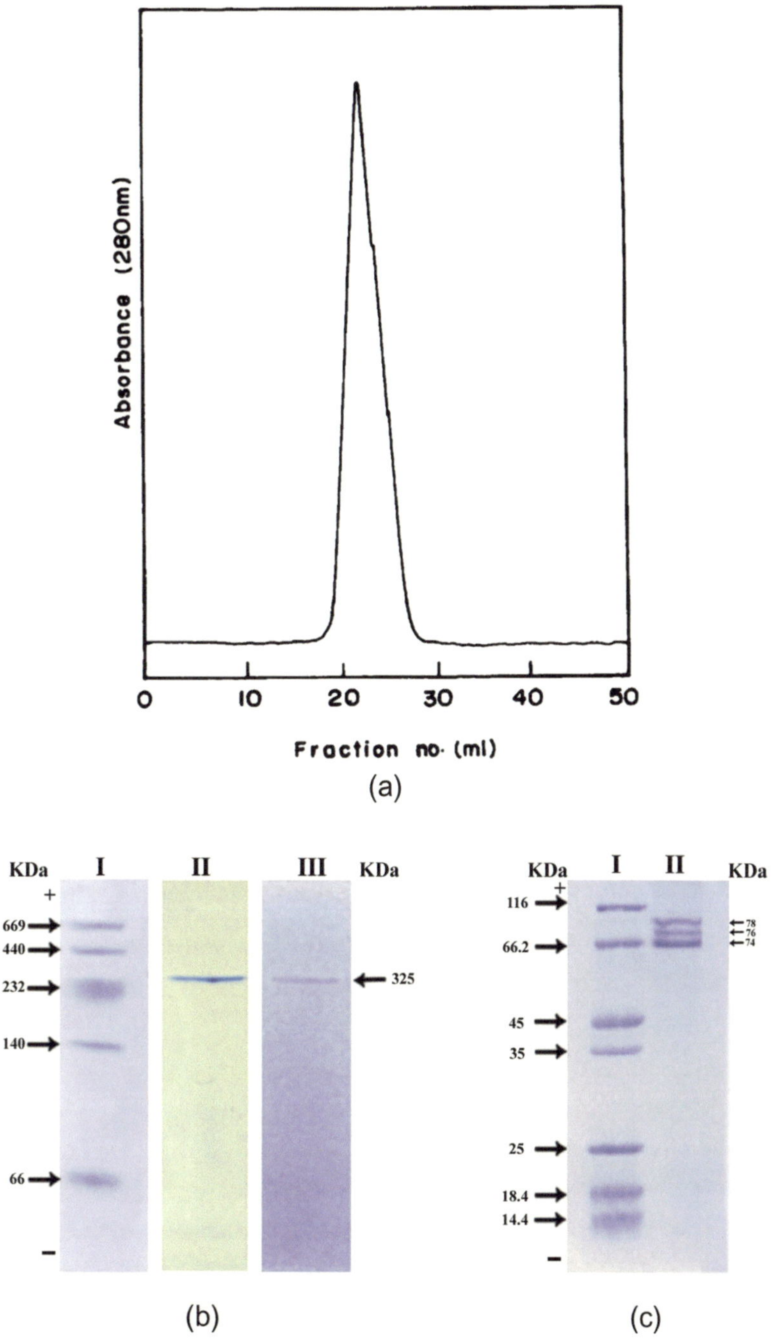

Figure 8.
(A) Gel filtration chromatographic profile of hemocyanin sample was applied on to the pre-equilibrated column of Sephadex G-100. The elution was performed at a flow rate 15 ml.h^{-1}. The fractions were continuously monitored for absorbance at 280 nm. (B) Electrophoretic analysis (PAGE—7%) of purified MrHC stained with CBB and BCSA after gel filtration chromatography from freshwater prawn M. rosenbergii. Lane 1: Molecular weight markers; lane 2: Purified MrHC (325 kDa) CBB stained; lane 3: Purified MrHC (325 kDa) bathocuproine sulfonic acid stained under UV light for copper-protein. (C) Electrophoretic profile of purified MrHC (325 kDa) protein was run under reduced conditions of sodium dodecyl sulfate–polyacrylamide gel electrophoresis (SDS-PAGE 10%) and stained with CBB. Lane 1 molecular weight protein markers; lane 2 purified MrHC (74, 76 and 78 kDa).

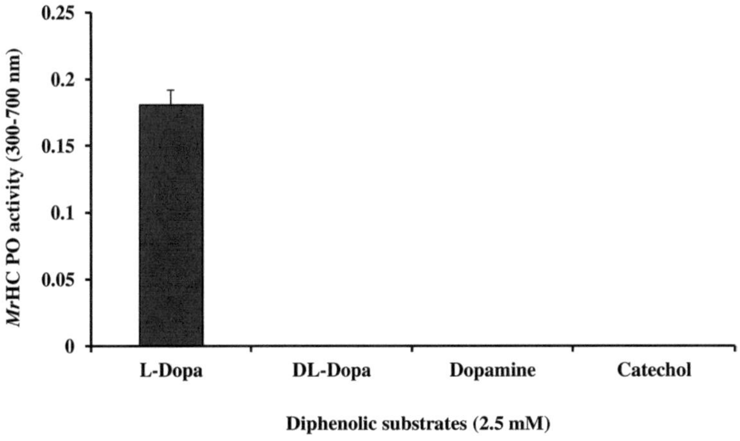

Figure 9.
PO activity of purified MrHC (325 kDa) with diphenolic substrates (2.5 mM) in Tris buffer (10 mM, pH 7.5) incubated at 25°C for 15 min and absorbance at 300–700 nm. The PO activity in optical density obtained at absorbancy maxima of respective substrates. Data represents mean ± SD of 5 determinations using purified MrHC (325 kDa) sample from different preparations.

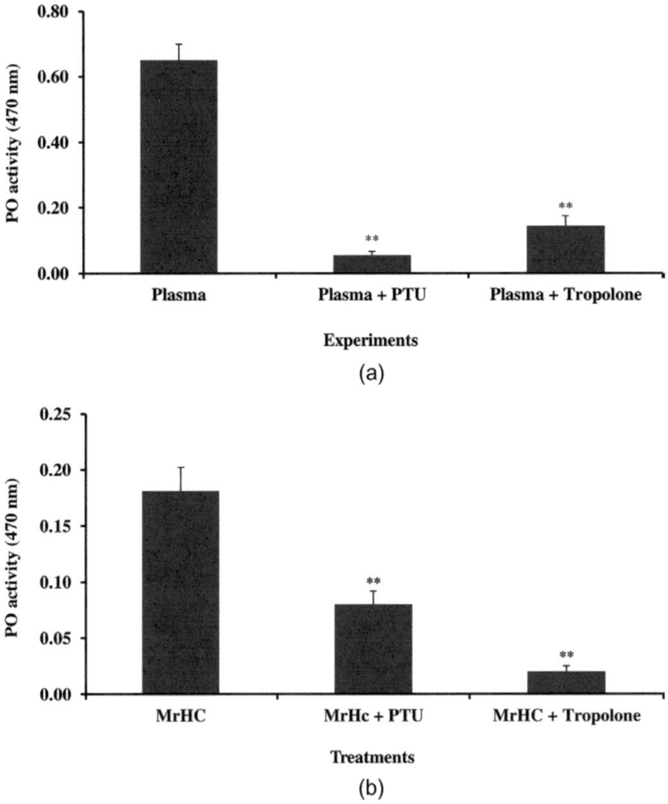

Figure 10.
*(A) Phenoloxidase activity in plasma and (B) purified MrHc (325 kDa) of M. rosenbergii and conformation of PO activity using inhibitors (PTU and tropolone). Asterisk indicates significant variation determined from the value obtained for a treatment to untreated control by one way ANOVA at **p < 0.01. Data represents mean of triplicate repeats of five determinations (mean ± SD).*

about 91.55% and 55.80%, respectively. However, pretreatment of plasma or purified *Mr*HC with tropolone (16 mM), another strong inhibitor of PO activity, also showed a significant reduction in the oxidation of L-DOPA when compared with control, and the reduction in the phenoloxidase with plasma and *Mr*Hc (77.88% and 88.95%) was statistically significant ($p < 0.01$) as shown in **Figures 10A** and **B**.

In summary, for the plasma or purified *Mr*HC (325 kDa), the optimal conditions for measuring PO activity on L-DOPA (2.5 mM) were 10 mM Tris–HCl, pH 7.5 at 25°C for 15 min at 470 nm.

4. Discussion

The hemocyanin showed phenoloxidase (PO) activity in *M. rosenbergii* and appears to be enhanced with activators such as proteases, SDS, and microbes [8] and agglutination activity [9]. The PO enzymes and hemocyanin molecules belong to the same class of copper proteins, and this explains its PO function [33]. Considering the complexity of crustacean immune defenses, our study attempted to explain the variance in the immune function of PO activity in the plasma and hemocyanin of *M. rosenbergii*. In most crustacean species active PO is a bifunctional enzyme that catalyzes *o*-hydroxylation of monophenols to diphenols and then oxidizes *o*-diphenols into *o*-quinone [15, 26]. The substrate affinity of plasma PO activity was attempted with monophenols including tyrosine and tyramine, diphenols such as L-DOPA, DL-DOPA, dopamine, catechol, or polyphenols such as hydroquinone and pyrogallol. The results clearly suggested that the plasma showed the highest substrate affinity with diphenols, and among the diphenols, L-DOPA was found to show the highest PO activity suggesting catechol oxidase activity.

Biochemical studies were undertaken to describe the optimum condition of the plasma PO activity. The enzyme reaction was observed with different concentrations of L-DOPA. There was a steady increase in the enzyme activity from 1 mM to 2.5 mM concentration of L-DOPA after which an increase in substrate concentration did not enhance the enzyme activity proving substrate inhibition as the cause of the decline in enzyme activity. The previous reported substrate-specific phenoloxidase activity of hemocytes derived from *Penaeus monodon* and *M. rosenbergii* using 1.6 mg/ml of L-DOPA [4] and *M. rosenbergii* injected with Gram-positive *Lactococcus garvieae* and Gram-negative *Aeromonas veronii* was monitored for changes in phenoloxidase (PO) activity using 10 mM L–DOPA [34].

Since PO is an enzyme, its activity depends on the steady state of the active sites, which are necessary for substrate binding and subsequent activity. The optimum ionic interactions were studied by taking the plasma in different ionic strength of Tris–HCl buffer, and PO activity was determined. The optimum ionic strength of 10 mM Tris–HCl that showed highest PO activity was used as a buffer for the study. To continue on ionic interactions, the optimum pH of the buffer required for plasma PO activity was also determined. The optimum pH was observed at pH 7.5 (brown color formation of dopachrome), which was same as that of purified *Charybdis japonica* PO [27] against L-DOPA and *Penaeus chinensis* [35], but different from that of brown shrimp *Penaeus californiensis* that showed optimum pH at 8.0 [36], *Penaeus setiferus* at pH 7.5 [37]. The differences in optimum pH may be correlated with the species specificity.

Temperature is an important factor that can either enhance enzyme activity or decline it. As the enzyme is a protein catalyst, a steady state of an active site binding to substrate depends on the intactness of the active site, which can be disrupted by temperature. In the present study, the optimum temperature of plasma PO activity in *M. rosenbergii* was determined by incubating plasma at various temperatures ranging from 10 to 90°C. The optimum temperature was found to be 25°C. The hemocyte of *M. rosenbergii* showed optimum PO activity at 37°C [4]. The differences in temperature optima in plasma and hemocytes suggest a difference in PO characteristics.

However, in different crustaceans, several authors found maximum activities of PO activity in a temperature range of 40–45°C [27, 35, 37–41] while reported maxima at 30°C and 55°C for shrimp *P. paulensis*, lobster *Homarus americanus,* and tiger prawn *P. monodon*, respectively [42–44]. The difference in temperature optima among different species of crustacea can be attributed to species specificity and existing physiological conditions. Time is an important factor that can either enhance PO enzyme activity or decline it. In the present study, the oxidation of the substrate (L-DOPA) was tested at various time intervals from 5 to 30 minutes, and the activity was observed to be high at 15 min of incubation. The hemocyte of *M. rosenbergii* showed optimum PO activity at 1 minute [4] while 40 minutes was recorded in *C. japonica* [27]. The differences in time interval optima in plasma and hemocytes suggest a difference in PO characteristics.

The enzyme kinetics of the plasma PO activity was determined using Michaelis–Menten curve by plotting various concentrations of L-DOPA (1–10 mM), and the rate of reaction was determined in 15 min (1/V). The initial rate of reaction increased up to a maximum reaction velocity after which it stabilized and then declined. The Km value determined for substrate enzyme affinity was 0.75 mM, and this suggested a strong affinity between the enzyme and L-DOPA and the Vmax was calculated as 0.58. Lineweaver-Burk plot showed a slope of 1.2 with a correlation coefficient of $R^2 = 0.996$. This indicated that the enzyme had active sites to maintain a steady increase in the rate of reaction. The kinetic and biochemical characteristics of the plasma PO activity demonstrate a distinct PO activity among the crustaceans [27, 45].

Our study also included the determination of PO activity concerning substrate affinity and inhibition using the optimized conditions as determined in plasma for hemocyanin (325 kDa) separated from the hemolymph of *M. rosenbergii*. This study on substrate affinity of purified *Mr*HC was undertaken with diphenols such as L-DOPA, DL-DOPA, dopamine, and catechol and was specifically PO activity with L-DOPA only. It failed to show any binding affinity with any other diphenolic groups, and this indicated its distinct catechol oxidase nature [46].

Comparative inhibition studies with the PTU and tropolone were made to confirm the PO activity in the plasma and purified *Mr*HC. The typical *o*-diphenoloxidase inhibitor, phenylthiourea, inhibited the enzyme activity drastically in the plasma and the tropolone inhibited the phenoloxidase activity in purified *Mr*HC. The inhibition studies revealed that the plasma and the purified *Mr*HC showed phenoloxidase activity. These results are following phenoloxidase from *P. californiensis* [36], *P. chinensis* [35], *C. japonica* (Liu et al. 2006) [27], and *Limulus polyphemus* [47]. Phenylthiourea (PTU), known as a chelating reagent of copper [47, 48], effectively inhibited the activity of plasma phenoloxidase activity and also that of purified *Mr*HC, suggesting that phenoloxidase from *M. rosenbergii* prawn has copper in its active site. Furthermore, the observed oxidation of L-DOPA was not due to peroxidase since tropolone that inhibited PO activity in the plasma and purified *Mr*HC did not act as a substrate for peroxidase in the presence of H_2O_2 [49, 50].

5. Conclusion

In the present study, we conclude that the immunological function of phenoloxidase observed in plasma and *Mr*HC (326 kDa) of freshwater prawn *M. rosenbergii* appears to enhance resistance against various diseases, and investigation of PO activity in plasma and hemocyanin protein revealed catechol oxidase type. However, for the plasma and purified *Mr*HC (325 kDa), the optimal conditions for measuring PO activity on L-DOPA (2.5 mM) were 10 mM Tris–HCl, pH 7.5 at 25°C for 15 min at 470 nm. This clearly indicates the significance of humoral immune components in boosting immune response. This finding provides evidence that the plasma and *Mr*HC of *M. rosenbergii* are a potent immune system with an ability to enzymatically function as humoral PO.

Funding

Not applicable.

Conflict of interest

The authors declare no conflict of interest.

Declarations

I confirm that the manuscript, or its contents in some other form, has not been published previously by any of the authors and/or is not under consideration for publication in another journal at the time of submission.

Abbreviations

CBB	Coomassie brilliant blue
DL–DOPA	3,4-dihydroxy-DL-phenylalanine
L–DOPA	L-3,4-dihydroxyphenylalanine
*Mr*HC	*M. rosenbergii* hemocyanin
PAGE	Polyacrylamide gel electrophoresis
PO	Phenoloxidase
*pro*PO	prophenoloxidase
PTU	Phenylthiourea
SDS–PAGE	Sodium dodecyl sulfate–polyacrylamide gel electrophoresis
TBS	Tris-buffered saline

Author details

Mullaivanam Ramasamy Sivakumar[1] and Rangasamy Shanthi[2]*

1 Laboratory of Pathobiology, Department of Zoology, University of Madras, Guindy Campus, Chennai, Tamil Nadu, India

2 Laboratory of Nutrition and Crustacean Biology, Department of Zoology, University of Madras, Guindy Campus, Chennai, Tamil Nadu, India

*Address all correspondence to: crusshanthi@gmail.com

References

[1] Avnimelech Y, Verdegem MCJ, Kurup M, Keshavanath P. Sustainable land–based aquaculture: Rational utilization of water, land and feed resources. Mediterranean Aquaculture Journal. 2008;**1**:45-55

[2] Jayasree L, Janakiram P, Madhavi R. Shell disease in the freshwater prawn *Macrobrachium rosenbergii* (de man): Aetiology: Pathogenicity and antibiotic sensitivity. Journal of Aquaculture in the Tropics. 1999;**14**:289-298

[3] Shailender M, Krishna PV, Suresh Babu CH, Srikanth B. Impact of diseases on the growth and survival of giant freshwater prawn, *Macrobrachium rosenbergii* (De man) larvae in the hatchery level. World Journal of Fish and Marine Sciences. 2012;**4**:620-625

[4] Sung HH, Chang HJ, Her CH, Chang JC, Song YL. Phenoloxidase activity of hemocytes derived from *Penaeus monodon* and *Macrobrachium rosenbergii*. Journal of Invertebrate Pathology. 1998;**71**:26-33

[5] Cheng W, Chen JC. Effects of pH, temperature and salinity on immune parameters of the freshwater prawn *Macrobrachium rosenbergii*. Fish & Shellfish Immunology. 2000;**10**:387-391

[6] Sritunyalucksana K, Sithisarn P, Withayachumnarnkul B, Flegel TW. A preliminary study on, activation of agglutinin and antibacterial activity in hemolymph of the black tiger prawn, *Penaeus monodon*. Fish & Shellfish Immunology. 1999;**9**:21-30

[7] Acharya S, Mohanty J, Sahoo PK. Humoral defense factors in Indian River prawn, *Macrobrachium malcolmsonii*. Fish & Shellfish Immunology. 2004;**17**:137-147

[8] Sivakumar MR, Denis M, Shanthi S, Arumugam M. Phenoloxidase activity in humoral plasma, hemocyanin and hemocyanin separated proteins of the giant freshwater prawn *Macrobrachium rosenbergii*. International Journal of Biological Macromolecules. 2017;**102**:977-985

[9] Sivakumar MR, Denis M, Shanthi S, Arumugam M. Agglutination of plasma, hemocyanin, and separated hemocyanin from the hemolymph of the freshwater prawn *Macrobrachium rosenbergii* (De man, 1879) (Decapoda: Caridea: Palaemonidae). Journal of Crustacean Biology. 2020;**40**:309-315

[10] Cerenius L, Söderhäll K. The prophenoloxidase-activating system in invertebrate. Immunological Reviews. 2004;**198**:116-126

[11] Liu H, Jiravanichpaisal P, Cerenius L, Lee BL, Söderhäll I, Söderhäll K. Phenoloxidase is an important component of the defense against *Aeromonas hydrophila* infection in a crustacean *Pacifastacus leniusculus*. The Journal of Biological Chemistry. 2007;**282**:33593-33598

[12] Kanost MR, Gorman MJ. Phenoloxidases in insect immunity. In: Beckage NE, editor. Insect Immunology. Amsterdam: Elsevier; 2008. pp. 69-96p

[13] Yamamoto K, Yakiyama M, Fujii H, Kusakabe T, Koga K, Aso Y, et al. Expression of prophenoloxidase mRNA during silkworm hemocyte development. Bioscience, Biotechnology, and Biochemistry. 2000;**64**:1197-1202

[14] Bai G, Brown JF, Watson C, Yoshino TP. Isolation and characterization of phenoloxidase from egg masses of

the gastropod mollusc, *Biomphalaria glabrata*. Comparative Biochemistry and Physiology Part B. 1997;**118**:463-469

[15] Sritunyalucksana K, Söderhäll K. The *pro*PO and clotting system in crustaceans. Aquaculture. 2000;**191**:53-69

[16] Fang J, Han Q, Johnson JK, Christensen BM, Li J. Functional expression and characterization of *Aedes aegypti* dopachrome conversion enzyme. Biochemical and Biophysical Research Communications. 2002;**290**:287-293

[17] Sugumaran M. Role of insect cuticle in immunity. In: Söderhäll K, Iwanaga S, Vasta G, editors. New Directions in Invertebrate Immunology. Fair Haven, NJ: SOS Publications; 1996. pp. 355-374

[18] Shelby KS, Adeyeye OA, Okot-Kotber BM, Webb BA. Parasitism–linked block of host plasma melanization. Journal of Invertebrate Pathology. 2000;**75**:218-225

[19] Olivares C, Jimenez-Cervantes C, Lozano JA, Solano F, Garcia-Borron JC. The 5,6–dihydroxyindole–2–carboxylic acid (DHICA) oxidase activity of human tyrosinase. Biochemical Journal. 2001;**354**:131-139

[20] Liu S, Zheng SC, Li YL, Li J, Liu HP. Hemocyte-mediated phagocytosis in crustaceans. Frontiers in Immunology. 2020;**11**:268

[21] Hernández-López J, Gollas-Galván T, Góumez-Jiméunez S, Portillo-Clark G, Vargas-Albores F. In the spiny lobster (*Panulirus interruptus*) the prophenoloxidase is located in plasma not in haemocytes. Fish & Shellfish Immunology. 2003;**14**:105-114

[22] Jaenicke E, Decker H. Tyrosinases from crustaceans form hexamers. Biochemical Journal. 2003;**371**:515-523

[23] Adachi K, Hirata T, Nishioka T, Sakaguchi M. Hemocyte components in crustaceans convert hemocyanin into a phenoloxidase–like enzyme. Comparative Biochemistry and Physiology Part B. 2003;**134**:135-141

[24] Coates CJ, Nairn J. Hemocyanin-derived phenoloxidase activity: A contributing factor to hyperpigmentation in *Nephrops norvegicus*. Food Chemistry. 2013;**140**:361-369

[25] Chen JS, Rolle RS, Marshall MR, Wei CI. Comparison of Phenoloxidase activity from Florida spiny lobster and Western Australian lobster. Journal of Food Science. 1991;**56**:154-157

[26] Decker H, Jaenicke E. Recent findings on phenoloxidase activity and antimicrobial activity of hemocyanins. Developmental and Comparative Immunology. 2004;**28**:673-678

[27] Liu G, Yang L, Fan T, Cong R, Tang Z, Sun W, et al. Purification and characterization of phenoloxidase from crab *Charybdis japonica*. Fish & Shellfish Immunology. 2006;**20**:47-57

[28] Coates CJ, Nairn J. Diverse immune functions of hemocyanins. Developmental and Comparative Immunology. 2014;**45**:43-55

[29] Bradford MM. A rapid and sensitive method for the quantitation of microgram quantities of protein utilizing the principle of protein dye binding. Analytical Biochemistry. 1976;**72**:248-254

[30] Maurer HR. Disc Electrophoresis and Related Techniques of Polyacrylamide Gel Electrophoresis. Berlin, New York: Walter de Gruyter; 1971. p. 222

[31] Bruyninckx WJ, Gutteridge S, Mason HS. Detection of copper on polyacrylamide gels. Analytical Biochemistry. 1978;**89**:174-177

[32] Laemmli UK. Cleavage of structural proteins during the assembly of the head of bacteriophage T4. Nature. 1970;**227**: 680-685

[33] Decker H, Terwilliger N. Cops and robbers: Putative evolution of copper oxygen-binding proteins. Journal of Experimental Biology. 2000;**203**:1777-1782

[34] Sung HH, Huang YT, Hsiao LT. Phenoloxidase activity of *Macrobrachium rosenbergii* after challenge with two kinds of pathogens: *Lactococcus garvieae* and *Aeromonas veronii*. Fish Pathology. 2004;**39**:1-8

[35] Fan TJ, Wang XF. Purification and partial biochemical characterization of phenoloxidase from *Penaeus chinensis*. Acta Biochimica et Biophysica Sinica. 2002;**34**:589-594

[36] Gollas-Galván T, Hernández-López J, Vargas-Albores F. Prophenoloxidase from brown shrimp *Penaeus californiensis* hemocytes. Comparative Biochemistry and Physiology Part B. 1999;**122**:77-82

[37] Simpson BK, Marshall MR, Otwell WS. Phenoloxidase from shrimp (*Penaeus setiferus*): Purification and some properties. Journal of Agricultural and Food Chemistry. 1987;**35**:918-921

[38] Simpson BK, Marshall MR, Otwell WS. Phenoloxidases from pink and white shrimp: Kinetic and other properties. Journal of Food Biochemistry. 1988;**12**:205-217

[39] Rolle RS, Guizani N, Chen JS, Marshall MR, Yang JS, Wei CI. Purification and characterization of phenoloxidase isoforms from Taiwanese black tiger shrimp (*Penaeus monodon*). Journal of Food Biochemistry. 1991;**15**:17-32

[40] Ali MT, Gleeson RA, Wei CI, Marshall MR. Activation mechanisms of pro–phenoloxidase on melanosis development in Florida spiny lobster (*Panulirus argus*) cuticle. Journal of Food Science. 1994;**59**:1024-1030

[41] Adachi K, Hirata T, Nagai K, Fujisawa S, Sakaguchi M. Purification and characterization of prophenoloxidase from Kuruma prawn *Penaeus japonicus*. Fisheries Science. 1999;**65**:919-925

[42] Opoku-Gyamfua A, Simpson BK, Squires EJ. Comparative studies of the polyphenoloxidase fraction from lobster and tyrosinase. Journal of Agricultural and Food Chemistry. 1992;**40**:772-775

[43] Perazzolo LM, Barracco MA. The prophenoloxidase activating system of the shrimp *Penaeus paulensis* and associated factors. Developmental and Comparative Immunology. 1997;**21**:385-395

[44] Montero MP, Avalos A, Pérez-Mateos M. Characterization of polyphenoloxidase of prawn (*Penaeus japonicus*). Alternatives to inhibition: Additives and high pressure treatment. Food Chemistry. 2001;**75**:317-324

[45] Jaenicke E, Decker H. Kinetic properties of catecholoxidase activity of tarantula hemocyanin. The FEBS Journal. 2008;**275**:1518-1528

[46] Quinn EA, Malkin SH, Rowley AF, Coates CJ. Laccase and catecholoxidase activities contribute to innate immunity in slipper limpets, *Crepidula fornicata*. Developmental and Comparative Immunology. 2020;**110**:103724

[47] Wright J, McCaskill CW, Cain JA, Patterson A, Coates CJ, Nairn J. Effects of known phenoloxidase inhibitors on hemocyanin-derived phenoloxidase

from *Limulus polyphemus*. Comparative Biochemistry and Physiology Part B. 2012;**163**:303-308

[48] Shanthi S, Sivakumar MR, Rayvathy B. Serum phenoloxidase activity in the hemolymph of the anomuran crab *Albunea symmysta* (Linnaeus, 1758) (Decapoda: Anomura: Albuneidae). Journal of Crustacean Biology. 2021;**41**:1-8

[49] Kahn V. Tropolone–a compound that can aid in differentiating between tyrosinase and peroxidase. Phytochemistry. 1985;**24**:915-920

[50] Brivio MF, Mazzei C, Scarf G. proPO system of *Allogamus auricollis* (Insecta): Effects of various compounds on phenoloxidase activity. Comparative Biochemistry and Physiology Part B. 1996;**113**:281-287

Chapter 7

Isolated Crayfish Stretch Receptor Neuron Electrophysiology May Explain a Longstanding Mystery of Human Brain Functioning: Eureka Moment

Toru Yazawa

Abstract

Neural network of our brain is complex, but single-neuron physiology is still important to understand the higher brain function. While conducting electrophysiological experiments using the isolated crayfish stretch receptor neuron, a phenomenon which may explain a longstanding mystery of human brain functioning, Eureka moment, was found. In this article, we demonstrate electro-physiologically GABAergic inhibitory synapses contribute for "switching" and propose a novel idea that can explain how sudden switching occurs in the brain.

Keywords: crayfish, stretch receptor neuron, GABA, principal cell, inhibitory interneuron, shunting inhibition, higher level, science of the mental

1. Introduction

"It's no secret that the brain's complexity is vast, and although scientists have come closer to understanding how it functions, there is still a long way to go [1]." Historically speaking [2], observed detailed architecture of the brain cells with silver staining (Golgi's method). Eccles [3] and Kuffler [4] illuminated the working of the nervous system on a large scale. Buzsáki [5] explained why the rhythms of the brain are important, and Stuart and Spruston [6] emphasized the critical role of dendrites in information processing in the brain.

Buzsáki and da Silva [7] mention that "high frequency oscillations constitute a novel trend in neurophysiology." Thus, oscillations, including fluctuation, synchronization, and coupling, of assembled neurons are distinct physiological events occurring in the brain. However, there seems to be no satisfactory explanation for the science of the mind: "Eureka moment," and "Rubin's vase" or the "rabbit–duck illusion" which are bi-stable switching with ambiguous images or reversible figures. The switching is subjective, autonomic, or voluntary transformation occurring in adult brain.

We should go back again to the basics, observing a single-isolated neuron, asking whether "switching" occurs. Neuron is physical machinery composed of three parts: the dendrite, the cell body, and the axon. The commonality must be evolutionary consequence of convergent adaptation, from crustacean to human.

It is known that gamma-amino butyric acid (GABA) is the most important and dominant inhibitory neurotransmitter substance in the brain of human and crustacean too. However, its biochemical characteristics, inhibitory nature, and evidences for its synaptic release have been not established in human brain. But rather, crustacean neurons were used to show a significance of GABA molecule as the most important inhibitory neurotransmitter. Using crustacean nerve cells in 1950s, a German neurobiologist Florey, and Kuffler and his lab researchers; e.g., Otsuka and Kravicz [8, 9], contributed for identifying GABA as an inhibitory substance [10].

In mammalian brain, isolation of a single principal cell such as "pyramidal cell" is possible but not practical idea. In the brain slice preparation, it is inevitable that undetectable actions might affect neuron's response when artificially stimulating target neuron(s) in the slice. Dissociated and isolated culture neuron are not adequate specimen because it is not matured naturally.

We would like to use the crustacean stretch receptor neuron, which is the mechanosensory receptor neuron (RN) of the muscle receptor organ (MRO). The RN is one of the most thoroughly understood neurons studied by many researchers [11–31]. Using this material as a model of the principal neuron (such as the pyramidal cells in the cortex), we examine whether "switching" is observable at a single-cell level. We consider that sudden change of electric current flows in the neurons might be occurring at "Eureka moment."

It is known that, in mammalian brain, the principal neurons receive abundant inhibitory inputs from the interneurons such as Ramón y Cajal's "basket" neurons. This inhibition is mediated by chloride ions [4]. MRO neurons (i.e., RNs) also receive GABAergic inhibition which is also mediated by chloride ions [4].

In the present study, we focus on the inhibitory transmission around the principal neurons. More precisely, we examine the role of GABAergic inhibition in the microcircuit. We will discuss the functional significance of GABAergic neurons for explaining neurophysiological mechanisms of "Eureka moment."

2. Single-neuron experiments

2.1 Cell body matters

Figure 1 shows five examples for the loci of soma within the architecture of neuron. A shows the soma of RN in the MRO. This neuron extends its axon to the brain (abdominal ganglia). In the ganglia, the axon connects to the inhibitory interneuron (black neuron in **Figure 1A**). RN receives recurrent inhibitory influence (pink arrow in **Figure 1A**). B shows the soma of the principal neuron of vertebrate. Pyramidal cell is known to send axon collateral to the inhibitory interneuron (i.e., Cajal's basket cell, black neuron in **Figure 1B**). C shows the spinal sensory cell. D and E show representative neurons of crustaceans, mollusks, and insects. There are distinct two types in the brain: one is A and B, the other is C, D, and E. The latter case (C, D, and E) is understandable, but the former case is not. The soma of A and B sit on the main route. It is seemingly an obstacle. It might be disturbing the information flow. The soma loci (A and B) might have undiscovered physico-physiological functions contributing for "neural computation" or "information-processing" in the highly cognitive brain.

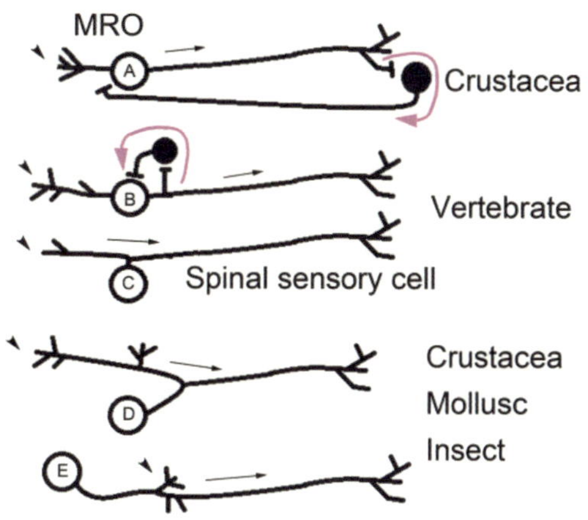

Figure 1.
Diagrammatic representation of neurons demonstrating the soma loci. A, crustacean MRO neuron. B, vertebrate principal neuron. C, vertebrate sensory neuron in the spinal cord. D and E, neurons of invertebrates such as crustaceans, mollusks, and insects. Arrow heads, afferent inputs. Arrows, direction of electrical signal flow. Pink arrows, recurrent inhibitory microcircuit.

Intriguingly and surprisingly, the soma A and B are covered by highly packed synaptic knobs on surface (see Figures 1–4, page 7 of ref. [3]). Most of synapses on the soma are inhibitory [3]. What is the function of the soma? Is it not obstacle? What is the function of inhibitory synapses densely distributing on the somal and axonal membranes? We found answers in the present study. We found an important role of soma-axonal inhibitory synapses. It might be working everywhere across our brain.

2.2 Cell isolation for electrophysiological experiments

2.2.1 Nerve dissection: Microscopic technique

Figure 2 shows dissection procedure of abdominal MRO in the present study. **Figure 2A** shows the locus of MRO, at the fourth segment (blue), **Figure 2B** shows MRO from rostral segments, and **Figure 2C** shows MRO from caudal segments. Note that caudal MROs are stick together. **Figure 2D–F** illustrates dissection procedure, separating two MROs off using a forceps and a fine needle. In **Figure 2E**, RNs may not seriously be injured at this state. But in **Figure 2F**, desheathing manipulation of the soma-axon portion of RNs (red) is conducted although it is venturous procedure. As a consequence, caudal MRO is suitable for obtaining a successful desheathed preparation.

Figure 3 shows light-microscopic view of MRO in the present study. The photographs were taken by a handy camera through an eyepiece. Left picture in **Figure 3** corresponds to the diagrammatic representation shown in **Figure 2C**. Arrows point axons of two RNs. F denotes one of the clamp forceps clamping at the cut ends of the receptor muscles. Right picture in **Figure 3** corresponds to the diagrammatic representation shown in **Figure 2F**. Axon diameter is less than 10 μm.

2.2.2 Electrophysiology: Methods

The fast and slowly adapting stretch receptor organs with their receptor neurons, muscles, and efferent nerves attached were dissected out from the abdomen (**Figure 2**). Fundamental electrophysiological methods are previously mentioned [32]. Briefly, A

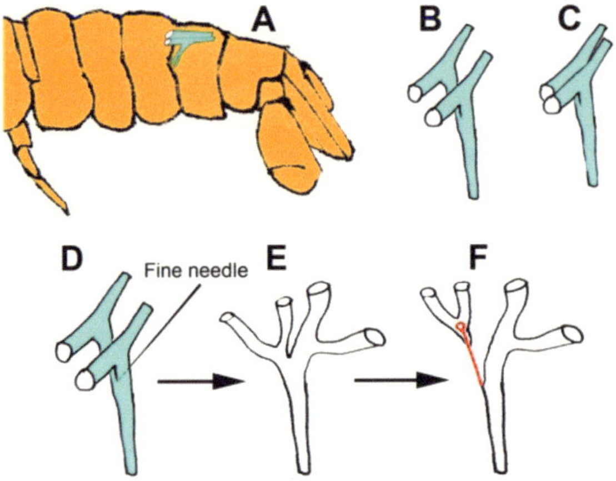

Figure 2.
Dissection procedure of abdominal MRO used in the present study. A, crayfish abdomen. B and C, isolated MROs from the abdomen. D–F, further dissection procedures after isolation.

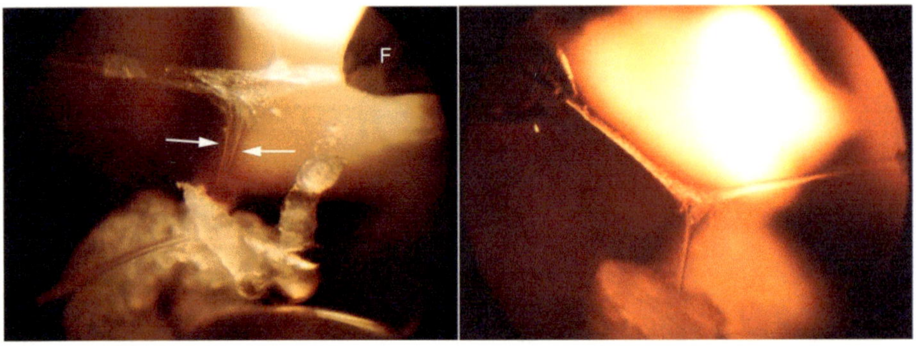

Figure 3.
Light-microscopic view of MRO. Left, an isolated MRO-RNs. Two axons of the slowly adapting and fast adapting RNs are pointed by arrows. Right, the slowly adapting neuron alone is isolated. One can see that an axon is exposed.

perfusion chamber contains 0.6 ml of van Harreveld solution [33], flowing speed 0.1 ml/s, flowing direction from dendrite to axon (see **Figure 4** inset), at room temperature, 18–22°C. Glass microelectrode is filled with 4 M potassium acetate, for recording and passing current through the membrane. For the iontophoretic application of GABA, a double-barreled electrode (see **Figure 4** inset) system was applied, one barrel filled with 1 M GABA, pH adjusted to 4.0 with HCl, and the other filled with 1 M NaCl. GABA was ejected as cations when a positive pulse was applied. For the prevention of possible leakage of GABA, negative current of 5 nA was continuously applied to the GABA electrode. The ejection current was superimposed on this breaking current. Stimulation of nerve was done by a suction electrode. The membrane potential and the iontophoretic current for GABA application were displayed on an oscilloscope and photographed. Facilities at the laboratory of Kazuo Ikeda (deceased 2016), Neuroscience, City Of Hope, Duarte, CA, USA, were used in 1985–1986. Crayfish *Procambarus clarkii* specimens were captured locally at Ontario, CA, USA.

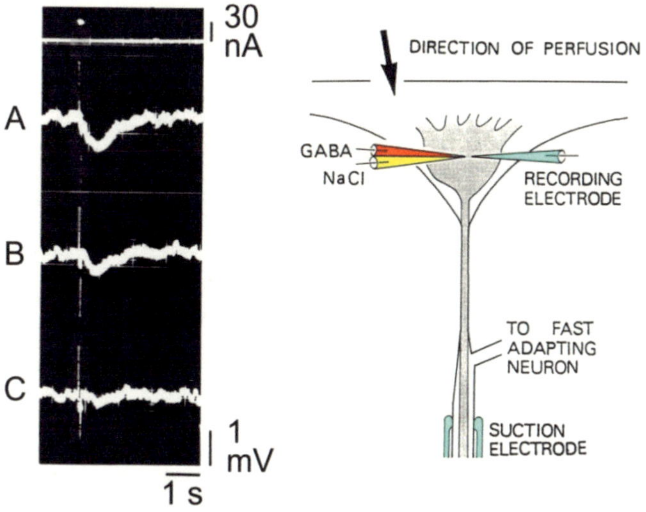

Figure 4.
GABA responses recorded intracellularly in the present study. Inset shows diagrammatically our experimental methods in general for the present article. GABA was applied to a spot on the soma iontophoretically by a 30 nA, 50 ms pulse. Uppermost trace shows the GABA current for A, B, and C. In A, the tip of GABA, electrode rested just at the surface of soma. In B, the GABA electrode was pulled back by 10 μm, and in C, by 20 μm.

2.2.3 Intracellular recording: Technical difficulty

We tried to reproduce recordings appeared in Kuffler's reports [15, 16, 21–23]. As shown in **Figure 5**, we successfully reproduced action potentials (APs) which were recorded from the soma membrane (**Figure 5B**). Inhibitory postsynaptic potentials (IPSPs) were also reproducible (**Figure 5B**). However, success rate is not high. Penetration of glass microelectrodes into the soma was an obstacle. For example, we obtained valuable data from only three neurons out of 37 neurons, during a two-week

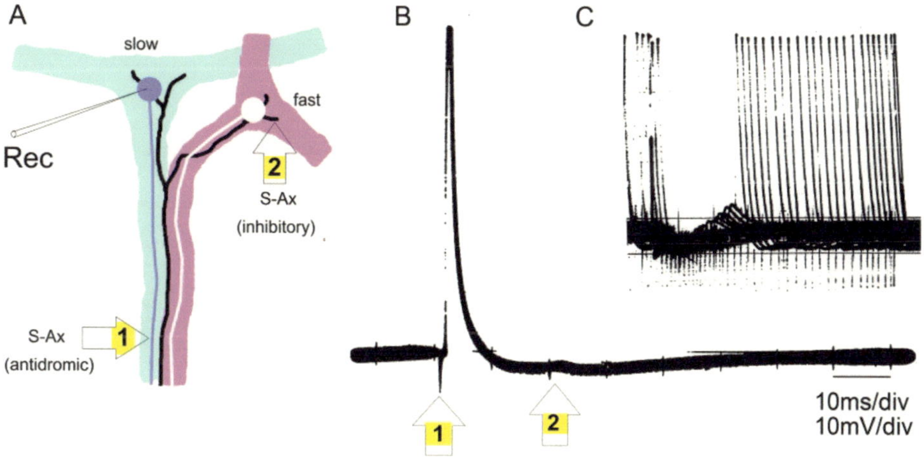

Figure 5.
We confirmed action potentials and inhibitory postsynaptic potentials in our preparation in the present study. Unpublished data.

	Wed 27Aug				Thu 28Aug				Fri 29Aug							
	Somite	Side	Cell	End	Somite	Side	Cell	End	Somite	Side	Cell	End				
1	4th	Right	Fast	Miss	4th	Right	Fast	Miss	2nd	Right	Fast	Miss				
2	4th	Left	Fast	Miss	4th	Left	Fast	Hit	2nd	Right	Fast	Miss				
3	3rd	Right	Fast	Miss	4th	Left	Fast	Miss	2nd	Left	Slow	Miss				
4	3rd	Right	Slow	Miss					2nd	Left	Fast	Miss				
5	3rd	Left	Fast	Miss					1st	Left	Fast	Miss				
6	2nd	Right	Fast	Miss					1st	Left	Slow	Miss				
7	2nd	Left	Fast	Miss					3rd	Left	Fast	Miss				
8									3rd	Right	Slow	Miss				
9									3rd	Right	Fast	Miss				
10									3rd	Right	Slow	Hit				
	Tue 2Sept				Wed 3Sept				Thu 4Sept				Fri 5Sept			
	Somite	Side	Cell	End	Somite	Side	Cell	End	Somite	Side	Cell	End	Somite	Side	Cell	End
1	2nd	Right	Fast	Miss	2nd	Right	Fast	Miss	2nd	Right	Fast	Miss	2nd	Right	Fast	Miss
2	2nd	Left	Fast	Miss	2nd	Right	Slow	Miss	2nd	Right	Slow	Miss	2nd	Right	Slow	Miss
3	3rd	Right	Fast	Miss	2nd	Left	Slow	Miss	3rd	Left	Slow	Miss	3rd	Left	Slow	Hit
4	3rd	Right	Slow	Miss	3rd	Right	Slow	Miss	2nd	Left	Slow	Miss				
5	3rd	Left	Slow	Miss												

Table 1.
Low success rate: A two-week five-day experiment.

five-day experiment, for example (see **Table 1**). **Table 1** summarizes what neuron we used and what end result we obtained, hit, or miss.

Figure 5A shows diagrammatical illustration of stimulation and recording methods in the present study. A glass microelectrode (Rec) is penetrated to the soma of neuron to observe membrane potential of slowly adapting neuron. S-Ax (arrow-1) denotes the point where antidromic stimulation of axon of slowly adapting receptor neuron (shown in blue-purple) was applied. Electrical pulse 0.05 ms duration was used at just above threshold intensity. S-Ax (arrow-2) denotes the point where stimulation of the inhibitory nerve (black) was applied. This inhibitory stimulation was done following the "axon reflex method" described by Kuffler and Eyzaguirre [23]. Using the just above threshold intensity is important, for preventing that excessive electric currents could stimulate other unrelated neuronal elements in the tissue.

Figure 5B shows antidromically initiated APs recorded at soma evoked by stimulation at S-Ax (arrow-1 in **Figure 5A**), and IPSPs evoked by stimulation at S-Ax (arrow-2 in **Figure 5A**). Time difference between arrow-1 and arrow-2 is fixed, approximately 20 ms, and several oscilloscope sweeps are superimposed.

Figure 5C shows superimposed weeps of APs and IPSPs. The time of antidromic stimulation was gradually shifted but IPSP stimulation time was fixed. IPSP blocks invasion of axonal spikes (failure of initiation of somal active spikes), but passive axonal spike is seen. B and C are different experiments with different cells thus, our experiments showing AP and IPSP are reproducible in the present study.

3. GABAergic inhibition

3.1 GABA iontophoretic application

Dendrites, soma, and axon areas receive a number of inhibitory synapse contacts, in both the MRO and the vertebrate principal neurons [3–6, 18–20]. This issue has not been well examined by physiological methods, because vertebrates, including human, neurons are impossible to isolate at a healthy-normal-operating state. Therefore, we considered that crustacean MRO neurons are beneficial. We examined synaptic function by iontophoretically applying small dose of GABA to localized area of the dendrites, soma, or axon (**Figures 4** and **6**). To do this experiment properly, it is imperative that the applied GABA is localized to a small spot so that there is no chance that other area will be stimulated by the application. Thus, the smallest detectable dose was determined, and the size of the affected area was estimated (**Figure 4**).

In **Figure 4**, we found that a single 30 nA current pulse of 50 ms duration produces a 1 mV response when the GABA electrode is located just on the surface of the somal membrane of the slowly adapting receptor. When the electrode was pulled back from the membrane, the latency of the response increased (see A and B in **Figure 4**) and the amplitude of the response diminished until it had disappeared completely with a 20 μm withdrawal from the membrane. This suggests that the GABA ejected by this current pulse is probably effective only in an area about 20 μm from the electrode.

When GABA current was increased, larger response with a similar time course was obtained (not shown). For the maximum GABA current employed in this experiment, the distance from the neuron surface to the tip of GABA electrode was 50 ± 10 μm to make the GABA response that disappears. Thus, even when the strongest ejection current (50 nA) was applied, synapses located more than 50 μm from the tip of the electrode were unaffected by the applied GABAs in this experiment.

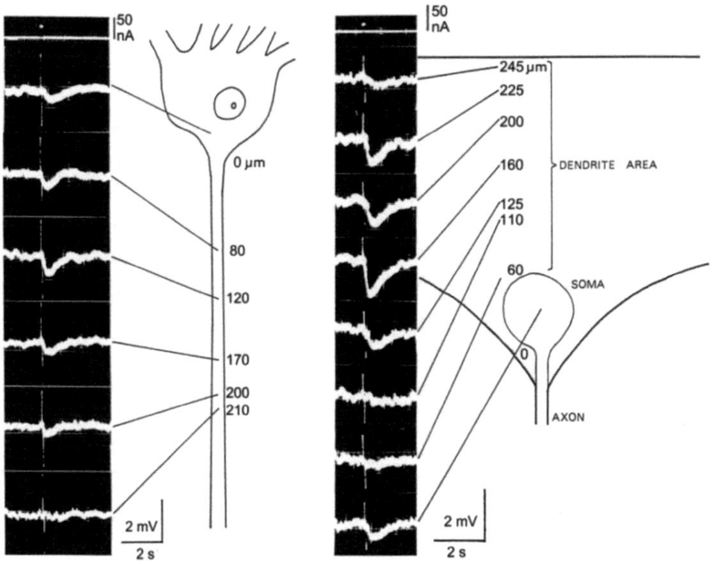

Figure 6.
Response to GABA applied locally on the various spots of the dendrite (right), and soma and axon (left). A 27 nA, 50 ms pulse was used for all GABA applications (uppermost trace). Application site was indicated in the diagram.

It is known that somal and axonal areas receive synapse contacts [18–20]. In **Figure 6**, we conducted the experiment surveying the GABA-sensitive area on the somal and axonal surfaces of the slowly adapting receptor neuron (**Figure 6** left panel). Hyperpolarizing responses can be seen when a small dose of GABA was iontophoretically applied to the soma and axon. The numbers in a diagram denote the distance from axon hillock. The experiment was repeated with 14 different preparations. Somal and axonal responses were obtained in every case.

A similar experiment was also done with an undesheathed preparation (likewise **Figure 2E**). When GABA was applied from the outside of the axon sheath, the response was almost null, even with the highest dose (50 nA, 1 sec).

With the same technique, the responses to GABA applied to the dendritic area were investigated with undesheathed preparations in the slowly adapting receptor neuron (**Figure 6** right panel). The amplitudes of the responses in the dendritic area were variable, depending on the location of the GABA electrode. The variability may well be related to the structure of the dendrite, but the fine structure of the dendrite in this recording condition was not clearly observable under a low-magnification microscope.

The electric currents using GABA application might cause direct effects on neuronal membrane. Thus, the possibility that the observed response was caused by the ejecting current, but not by GABA, was tested using a double-barreled electrode (see inset **Figure 4**) in which one barrel was filled with 1 M GABA and the other with 1 M NaCl (**Figure 7**). In this experiment, GABA was first applied to the dendrite, and then, the same current was passed through the NaCl barrel. As can be seen in **Figure 7A** and **B**, the neuron responded only to GABA.

It is known that neurotransmitter substances affect both pre- and postsynaptic terminals [4, 25]. There is a possibility that the response could be mediated through a presynaptic terminal. If the iontophoretically applied GABA was affecting inhibitory

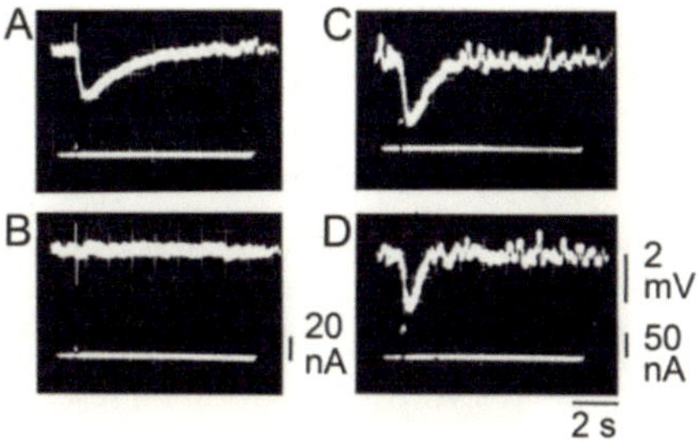

Figure 7.
A, response to GABA and B, that to NaCl applied to the same site in the dendritic area of the slowly adapting receptor neuron of the 3rd abdominal segment using double-barreled electrode (see inset in **Figure 4***). The same 12.5 nA iontophoretic current of 50 ms duration was used. C, D, response to GABA in a saline containing 3 mM $CoCl_2$.C, soma and D, axon (70 μm away from axon hillock).*

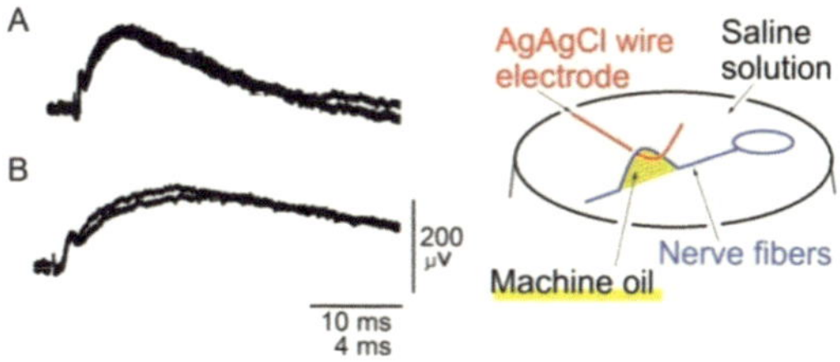

Figure 8.
IPSPs recorded extracellularly from a site on the axon between 90 μm and 180 μm from the soma of the slowly adapting receptor neuron of the 3rd abdominal segment. Inset shows diagrammatically experimental method (see also **Figure 5** *inset for the method of inhibitory stimulation). In both A and B, two CRO traces were superimposed. In the record with faster sweep speed (in B), an impulse of the inhibitory axon preceding the IPSP is seen (B is 2.5 times faster than A).*

terminals, which in turn released transmitter to the receptor neurons, an inhibitory potential would result. This possibility was tested by applying GABA in the presence of 3 mM $CoCl_2$ (see a standard text book, such as ref. [4]). The receptor neuron is responded to iontophoretically applied GABA in a 3 mM $CoCl_2$-containing solution in a similar manner to that in the standard van Harreveld solution (**Figure 7C** and **D**). It was, thus, confirmed that the applied GABA directly affects the postsynaptic membrane of the receptor neuron.

3.2 IPSP at the axon

The above results show that the GABA receptor exists not only on the dendrites, but also on the soma and the axon. Further evidence can be obtained by the focal recording of extracellular IPSPs upon stimulation of the inhibitory neuron. As shown in **Figure 8**, the local IPSP was recorded from the axon of slowly adapting receptor neuron at a portion of the axon 90–180 μm away from the soma. In the record (**Figure 8B**), the impulse of the inhibitory axon preceding the IPSP can be seen.

4. Function of the somal and axonal inhibitions

The stretch receptor neurons in crustaceans are one of the most extensively studied neurons [11–31]. The input signal, i.e., sensory-muscle extension, stimulates the dendrites and induces the generator potential at the dendritic membrane [4]. The depolarizing current conducts to the axon causing the initiation of the action potential [4]. The greater of the generator potential, the receptor neuron produces the higher frequency of discharge of action potential [4]. The inhibitory neuron can reduce the amount of the generator depolarization. This is the long-established understanding of neuronal function [4].

However, the above results in the present study revealed that crayfish stretch receptor neuron is innervated by the inhibitory neuron not only on its dendrite but also on its "soma" and part of its "axon." The presence of GABA-mediated inhibitory synapse is conclusively proven in the present study.

In our brain, the principal neuron (e.g., the pyramidal cell in the cerebral cortex) is innervated by the inhibitory neuron [3]. The most notable inhibitory interneurons that innervate the principal neuron are two cells: the basket cell and axo-axonic (or chandelier) cell [3–5]. The former innervates on its soma, and the latter innervates on its axon (i.e., initial segment or axon hillock). It is believed that the principal neuron plays important roles in advanced cognitive functions [6, 7].

Questions arise. What is the function of these inhibitory synapses distributing on the somal and axonal membranes? Do they work to attenuate the excitatory input? The functional role of the inhibitory synapses on the axon, axon hillock, or initial segment, should be reexamined.

4.1 Novel inhibitory function

4.1.1 Blockade of antidromic invasion of impulses into the soma by GABA applied to the axon

Axonal GABA receptors are demonstrated in **Figure 6** (see also [14]). However, the functional significance of axonal GABAergic synapse is not demonstrated physiologically [4, 5]. It is understandable that dendritic and somal inhibitory GABAergic synapses can inhibit neuronal activity. But, what is the functional significance of axonal GABAergic synapses? Why the axon receives GABAergic synapses?

We investigated the function of inhibitory synapses on the axon by observing nerve impulses with an intracellular electrode inserted into the soma (**Figure 9**). The cut end of the axon was stimulated while applying GABA iontophoretically onto a part of the axon 40 μm away from the soma. In the example shown in **Figure 9**, while the axon was being stimulated with a frequency of 2 Hz, GABA was being applied with a single ejecting current of 50 nA and 500 ms, a portion of the axon 40 μm away from the soma. Impulses designated A to F were recorded successively. GABA was applied for 50 msec beginning slightly before record B and ending prior to record C. In C and D, the somal regenerative impulses are blocked, leaving the axonal impulses, which electrically spread into the soma. The GABA effect lasted for about 1 sec. Thus, it is demonstrated that the invasion of the impulse into the soma was blocked, so that only the axonal impulse which is occurring prior to the point of inhibition by GABA is observed. We also observed that the blockage of the somal invasion required a large dose of GABA when applied to a single spot on the axon. In comparison to the dose of GABA which causes a detectable membrane hyperpolarization or depolarization (see **Figure 4**), the dose necessary for the blockage was more than 10 times higher. However, it can be assumed that this dose does not mean that the conduction block requires a higher dose of GABA to induce a larger conductance increase at a spot on the axon but rather indicates that a larger area of the axon must be inhibited to block

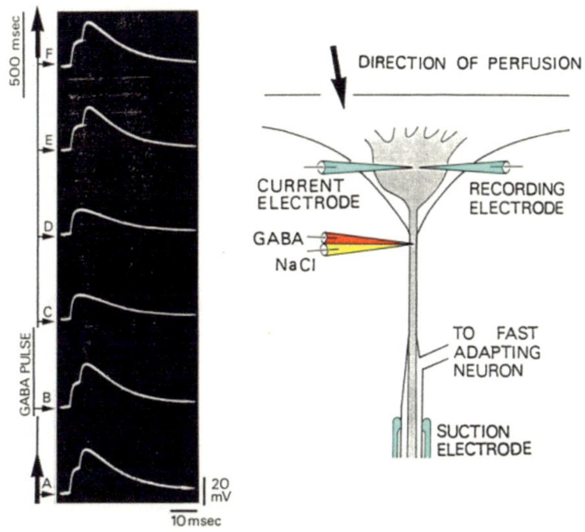

Figure 9.
Blockade of antidromic invasion of impulses into the soma by iontophoretic application of GABA onto the axon. Inset shows diagrammatically present experimental method. Antidromic impulses were elicited by stimulating (0.05 ms duration square pulse, blue-color suction electrode) the cut end of the axon of the slowly adapting stretch receptor neuron of the 3rd abdominal segment: Every 500 ms. GABA was applied with a pulse of 50 nA and 500 ms at a portion of the axon 40 nm away from the soma at the time indicated in the figure. Impulses were recorded successively on moving film. A shows the control before application of GABA. The GABA pulse was applied just before the impulse in B, and ended before C. One can see that somal invasion was blocked in C and D and recovered in E and F. In E, the conductance increase caused by GABA is seen as a faster time course of the impulse.

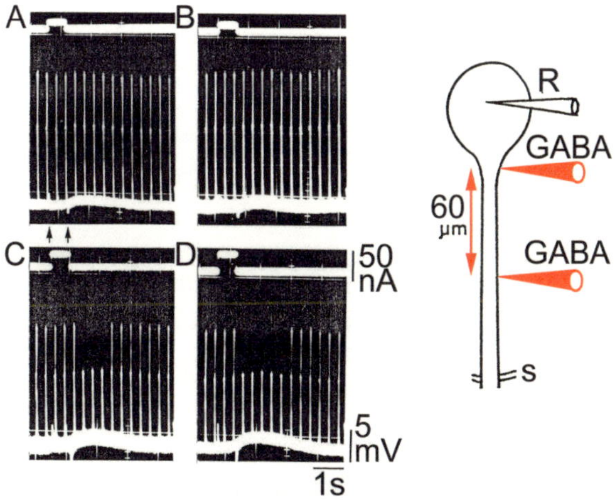

Figure 10.
Spatial effect of inhibition on the blockage of antidromic impulses. Inset diagrammatically presents experimental method. A: 9 nA, 500 ms pulse at 0 µm, No blockage. B: 9 nA, 500 ms pulse at 60 µm, No blockage. C: 9 nA, 500 ms simultaneous pulses at 0, 60 µm. Three impulses failed to invade. D: 12 nA, 500 ms simultaneous pulses at 0, 60 µm, five impulses failed to invade.

an impulse, i.e., the effectiveness is related to the spatial coverage rather than the concentration at a particular spot. Blockage of the conducting impulse may require a certain length of axon to be inactivated (i.e., shunt); otherwise, the impulse may jump the inactivated area. This possibility was tested below (see **Figure 10**).

In **Figure 10**, antidromic impulses were elicited with a frequency of 3 Hz by stimulating the axon of a slowly adapting receptor neuron of the 3rd abdominal segment. GABA was applied with two electrodes, one placed at the axon hillock (0 μm), the other 60 μm down the axon (see inset of **Figure 10**). Two glass capillary microelectrodes of the same resistance when filled with 1 M GABA were selected. Two electrodes in parallel were connected to a 1000 M ohm resister through which the ejecting current was applied.

Figure 10 shows that single-GABA electrode did not block antidromic spikes (see **Figure 10A** and **B**). But, two GABA electrodes, 60 μm apart each other, blocked antidromic spikes (see **Figure 10C** and **D**). Experiments shown in **Figure 10** demonstrate that blockage of the conducting impulse requires a certain length of axon to be inactivated (i.e., shunt); otherwise, the impulse may jump the inactivated area.

4.1.2 Effect of invading impulses on the stretch-induced impulse

Experiments shown in **Figure 10** demonstrated that antidromic-impulse invasion can be blocked when axonal GABAergic synapses are activated. Action potentials of neuron arise generally at the axon initial segment of neuron [4, 6]. Then, the action potentials travel toward two directions: one for orthodromic direction and the other for antidromic direction [4, 6]. It is obvious that orthodromic impulses convey neural information to the next cells. But functional significance of antidromic impulses is still controversial (see ref. [6, 34]).

In order to test the possible interaction between invading antidromic impulses and stretch-induced impulses, antidromic stimulation was given while the slowly adapting receptor neuron was stretched and held at constant tension. In **Figure 11A**, the impulse above the horizontal bar is antidromic impulses invading the soma. The stretch-induced impulses are seen to be accompanied by the generator potentials. As can be seen here, the frequency of the stretch-induced impulses becomes reduced for a certain period after the antidromic stimulation. In **Figure 11B**, the somal invasion was blocked for the entire period. Thus, all of the impulses seen here are axonal impulses recorded at the soma electrically. As can be seen, the antidromic stimulation

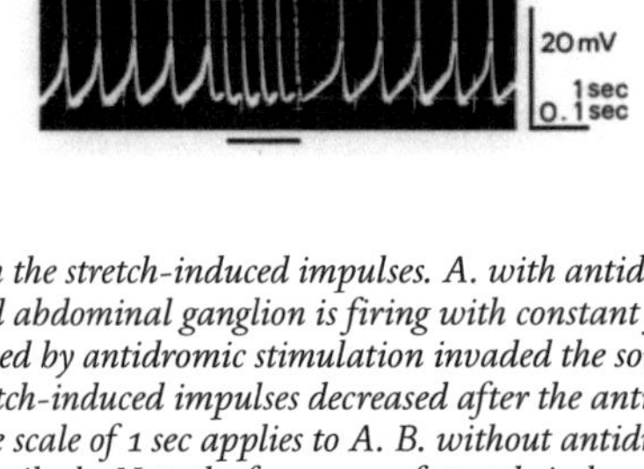

Figure 11.
Effect of antidromic stimulation on the stretch-induced impulses. A. with antidromic invasion. While a slowly adapting receptor neuron of the 3rd abdominal ganglion is firing with constant frequency with somal impulses unblocked, repetitive impulses caused by antidromic stimulation invaded the soma (12 impulses at 7 Hz above the bar). Note the frequency of stretch-induced impulses decreased after the antidromic stimulation, even though the tension was held constant. Time scale of 1 sec applies to A. B. without antidromic invasion. After blocking invasion, the effect was observed similarly. Note the frequency of stretch-induced impulses is unchanged before and after the antidromic stimulation (5 impulses above the bar). Time scale of 0.1 sec applies to B.

failed to affect the frequency of the stretch-induced impulses. This set of experiments demonstrates that when the somal-dendritic area is invaded by regenerative impulses, the frequency of the stretch-induced impulses is reduced. However, if the invasion is blocked, the frequency remains unchanged.

4.1.3 Effect of antidromic invasion of stretch-induced impulses on the stretch-induced frequency

When orthodromic impulses are set up by stretch, the excitation occurring at the spike initiation site also sends impulses back to the soma [4, 15, 22, 23]. This antidromic invasion must have an effect on the stretch-induced frequency. In other words, unless antidromic invasion is blocked, the output frequency will be lower than the frequency dictated by the generator potential, which is directly coupled with tension.

In order to demonstrate the effect of these returning impulses (antidromic invasion) on their own generator potential, constant tension was given to a slowly adapting receptor to elicit orthodromic impulses of constant frequency, and then GABA was applied to the axon in order to block antidromic invasion.

As can be seen in **Figure 12**, while the receptor neuron was firing at a constant frequency, GABA was applied to a spot on the axon 180 μm away from the soma. Antidromic invasion was blocked during the effective period indicated by the horizontal bar (16 axonal spikes without active somal membrane activities). It is clearly demonstrated that the frequency of firing during this period increased and stayed at a higher constant level. The frequency is returned to the original level after the cessation of the GABA action (**Figure 12**).

This result indicates that when antidromic invasion is allowed, the output frequency is lowered from the level originally set by the fixed tension. Only when antidromic invasion is blocked, does the output frequency precisely reflect the level of tension.

A question arises. How much difference is there between the precise level and the lowered (distorted) level? The frequencies of "higher constant level" and "original

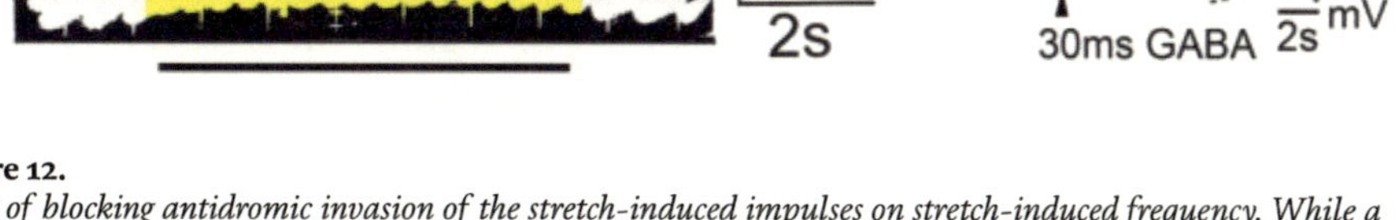

Figure 12.
Effect of blocking antidromic invasion of the stretch-induced impulses on stretch-induced frequency. While a slowly adapting receptor neuron of the 3rd abdominal segment was firing at constant frequency with its somal impulses unblocked, GABA was applied with a 50 nA, 2.5 sec pulse (top trace). During the period of GABA action (above the bar, colored yellow), antidromic invasion was blocked. Note the increased frequency of orthodromic impulses even though the tension was held constant. After GABA action, the frequency returned to the original level. Inset (right), a photograph from different specimens with a slower sweep speed. The receptor muscle was kept at a constant tension. Note constant frequency of impulse firing. A brief (30 ms duration, strong current) GABA application to the axon induced the blockage of the antidromic impulses. During the blockage period, it is clearly seen that hyperpolarizing after potential, which is mediated by potassium ions, is disappeared due to disappearance of the somal spikes.

constant level" are measurable from the oscilloscope-photographed data (9 cells). Seventeen measurements show following results, but it seems that data should not be statistically taken account because technical condition (dissection technique, electrode penetration technique, GABA electrode variation, etc.) varies.

Even though it is measurable in the present studies (all unpublished results). In **Figure 12**, high-level spiking discharge was at 2.47 Hz and low level at 1.64 Hz; thus, the invasion lowered the true information down to a 60 percent level, which is "false-distorted" information but meaningful information (see discussion). The most lowered case was 39 percent. The next significantly lowered case was 48 percent (3.33 Hz vs. 1.61 Hz). In **Figure 12** inset (right), the antidromic invasion lowers the information quality to the accuracy of 66 percent.

All other data obtained from the oscilloscope photographs are as follows: In the present studies, we obtained following calculated numbers: 60, 70, 71, 71, 59, 68, 57, 64, 57, 68, 61, 70, 77, and 78(in percent). In conclusion, the blockage can push the output frequency up to approximately 1.5 times.

5. Discussion

5.1 Crustacean stretch receptor cell

When the stretch receptor is uninhibited, antidromic impulse invades the soma. The present results demonstrate that the invasion is blocked when a part of axon is inhibited by the iontophoretic application of GABA. The effect of inhibitory postsynaptic potentials on the antidromically invading impulse was studied by Kuffler and Eyzaguire [23] by stimulating the inhibitory nerve while observing the response at the soma. They reported only a 2 percent decrease in the amplitude of the impulse and no blockage (see Figure 14 of ref. [23]). The inhibitory nerve synapses are not only on the soma and the axon, but also on the dendrites. Thus, stimulating the inhibitory nerve cannot differentiate the function of synapses localized on a particular part of the neuron. This is the reason why iontophoretic application of GABA was used in the present study in place of inhibitory nerve stimulation.

The failure to block the invasion of impulse to the soma in the experiment reported by Kuffler and Eyzaguirre [23] is understandable by looking at the record shown in that paper. The soma is depolarized by a large IPSP (see Figure 14 of ref. [23]), thus, providing a facilitative effect on the invading impulse. The effect of increased conductance must be overcome by this depolarizing potential change. When GABA was applied to a spot on the axon, the dose required for the blockage was as high as the equivalent of 2.5×10^{-8} coulombs in the example shown in **Figure 4**. When GABA was applied with two electrodes 60 μm apart along the axon, the necessary dose for blockage was lowered to the equivalent of 4.5×10^{-9} coulombs. In this case, the applied GABA had to diffuse to cover many synapses along the axon. It seems reasonable that the dose required for blockage by applying GABA at one spot was about five times higher than that required for the application with two electrodes separated from each other by a reasonable distance to effectively cover a certain length of axon. The length of axon necessary to effectively block the conducting impulse was, thus, estimated to be about 100 μm. In **Figure 4**, it was demonstrated that the dose of iontophoretically applied GABA on the axon necessary to cause a hyperpolarizing response of 1 mV was about 7.5×10^{-10} coulombs. If the transport number is assumed to be 0.5, this would correspond to 3.7×10^{-15} moles of GABA. In the neuromuscular junction of the same species of crayfish, Takeuchi and Takeuchi [35] reported that iontophoretic application of 4.2×10^{-15} moles of GABA elicited a depolarizing response of 0.5 mV. Of course, the polarity and the amplitude of the GABA response is dependent on the membrane potential relative to the equilibrium potential of the carrier ion, and deviation of these potentials from preparation to preparation but suggest that the sensitivity of axonal GABA receptors is similar to that of the muscle.

From the present experiment, the estimation of the axon length to be kept under inhibition in order to effectively block the conducting impulses was about 100 μm. The present experiment (**Figure 6**) showed that a part of the axon about 200–300 μm from the soma was found covered with inhibitory synapses and sensitive to GABA. It indicates that the distribution of inhibitory synapses along this length of axon provides a safety factor of 2–3 times for the blockage of conducting impulses.

The requirement that a certain length of axon must be under inhibition in order to block conducting impulses may explain why the orthodromic impulse initiation site of the neuron is at a location on the axon about 300 μm away from the soma. When constant tension is applied to the dendrites of the slowly adapting neuron by stretching the receptor muscle, the receptor neuron fires repetitively with a constant frequency determined by a steady generator potential level which is linearly related to the tension. These orthodromic impulses are initiated at a spike initiation site on the axon about 300 μm away from the soma, which is outside of the axonal area covered by inhibitory synapses. In the stretch receptor neuron of the lobster, the impulse initiation site is reported to be 500 μm away from the soma [16]. The impulses not only propagate toward the CNS (orthodromic direction), but also propagate back to the soma (antidromic). Thus, the antidromic impulse invading the soma is recordable with an intracellular electrode inserted into the soma.

In the experiment shown in **Figure 12**, the receptor neuron was kept under constant tension to induce firing of constant frequency. When GABA was applied to the axon at a spot 180 μm away from the soma, the invasion of impulses to the soma was blocked, leaving axonal impulse only. During this blocking period, the frequency of stretch-induced impulses increased, although the stretch had been kept constant. This implies that the higher frequency under the blockage of antidromic invasion of impulses is the frequency directly related to the generator potential level set by the steady tension. The frequency without blockage of antidromic invasion, on the other hand, is a frequency affected by the antidromic impulses. In other words, when antidromic invasion occurs, the frequency of impulses becomes lower than that directly set by the tension.

When the neuron is firing as a result of a given tension, the persisting generator potential keeps the membrane at a depolarized level set by the tension. If an impulse invades the soma in this condition, the repolarizing phase of the impulse creates a hyperpolarizing after potential which is superimposed on the generator potential—resulting in a lowering of the level of the generator potential. This will naturally reduce the frequency of firing. The effect of lowering the generator potential level by invading impulses must be dependent on the frequency, because the mechanism is apparently due to the summation of hyperpolarizing after potentials. Thus, the higher the level of the generator potential, the larger the hyperpolarizing after potential. Therefore, the effect of the invading impulse on the frequency is more stressed when the receptor is firing at a higher frequency. It is predicted that the frequency-tension relationship will be skewed from linearity at the high-frequency end.

Only if antidromic invasion of its own impulses is blocked then the frequency-tension relationship can become linear. This must be the role of the inhibitory synapse distributing on the soma and on the axon, between the soma and the spike initiation site. Orthodromic impulses traveling toward the CNS are not blocked by GABA applied to the soma or axon. Thus, under this inhibition, the CNS receives information without being modulated by antidromic invasion—information of high fidelity.

5.2 Vertebrate brain cell

Inhibitory synapses on the soma and axon hillock have been found in many other neurons (e.g., pyramidal cell). Most of physiological paper has interpreted the function of these synapses as being to attenuate the excitatory input—classical postsynaptic inhibition—the dendrites were originally thought to act as simple receivers (p. 1718 in ref. [6]). The present study has demonstrated that the function of the inhibitory synapse on the soma and the initial segment of the axon is not the classical one, but rather is to protect the dendritic area where input is received from disturbance coming from the backfiring of the soma caused by antidromic invasion of its own impulses. This thereby provides high fidelity for the input-output relationship of the neuron.

5.2.1 Axo-axonic cell or chandelier cell

As shown in the review [36], GABAergic chandelier cells (axo-axonic cells first documented by Ramón y Cajal) have a unique arborization (see Figure 2 of ref. [36]). One axo-axonic cell innervates 26 pyramidal cell's axon initial segment. Schneider-Menzel et al. [37] reported that 153 pyramidal cells receive chandelier input, although the synapse number is highly variable across the population and is correlated with structure features of the target neuron. This indicates that an interneuron connects to multiple principal cells. Furthermore, pyramidal cells are known to be electrically coupled. Single neurons and populations work together [38]. It is understandable that biophysically homogenous population [39] of neurons, assembly of neurons, can synchronously function as the switching device. We assume that this switch works as powerful (i.e., working simultaneously across the large brain area) converter of specialized cognitive machinery. "Eureka moment" and "rabbit-duck illusion" might use this switch, because we sometimes behave momentary extremely concentrated mode to think about just one thing like thoughtful Archimedes did.

5.2.2 GABA in the brain neural network

A review [40] described: "neural networks in the brain include principal neurons and GABAergic interneurons (e.g., basket cell). The latter is vital for normal brain function because they regulate the activity of principal neurons. PV (parvalbumin) interneurons (i.e., basket cell) are involved in gamma-frequency oscillations, and they also play a role in complex network operations, including expansion of dynamic activity range, pattern separation, modulation of place and grid field shapes, phase perception and gain modulation of sensory response. PV interneurons also play key role in numerous brain diseases. These include epilepsy and also complex psychiatric diseases such as schizophrenia. Thus, PV interneurons may become important therapeutic targets in the future."

The review continues: "However, much needs to be learned about the basic function of these interneurons before clinical neuroscientists will have a chance to successfully use PV neurons for therapeutic purpose." The present finding on the crustacean stretch receptor neurons provides a novel insight how PV interneurons shape complex network function.

Electrical stimulation to a special point of brain slice is usually not feasible because electric current cannot precisely stimulate the target, producing non-specific influences on many other unrelated neurons. Using GABA application to a point, Gómez et al. [41] studied somatodendritic inhibitory effects of GABAergic transmission. The GABAergic shunt mechanism supporting cognition could be proved in the brain one day. "The blockage of backfire at the axon initial segment" requires verification in the intact vertebrate brain.

5.2.3 GABA and chloride ion

Inhibitory postsynaptic potential (IPSP) at GABAergic synapse is mediated by chloride ionic currents. The reversal potential of IPSP (*E*i) is close to the resting membrane potential (*E*m). *E*m was −60 mV with a good condition in the present study. If *E*i is assumed to be −65 mV for example, IPSP must show a hyperpolarizing potential. As shown in **Figure 5C**, hyperpolarizing IPSP blocked antidromic invasion (cf. Figure 14 in ref. [23]). If *E*m is close to *E*i, GABA causes little change in membrane potential but a significant decrease in input resistance because chloride channels are open. This shunt inhibition is the mechanism of blockade of antidromic invasion. If *E*i is kept near *E*m, inhibitory nerve can induce only shunting effects without altering the membrane potential.

GABAergic signaling is controlled by the intracellular concentration of chloride ions. The chloride-extruding K^+-Cl^- cotransporter (KCC2) and the Na^+-K^+-$2Cl^-$ cotransporter (NKCC1) which facilitates the accumulation of Cl^- in neurons, maintain intracellular concentration of chloride ions. Thus, GABAergic signaling is controlled by both KCC2 and NKCC1. It is understandable that functional defect of them leads to the alteration in chloride homeostasis in CNS cells. GABA-chloride relationship is important for brain healthiness. Blockage of backfiring under the chloride ionic mechanism could be a therapeutic target in pathophysiology and pharmacology: the Renshaw cell in the spinal cord (this is not GABAergic but chloride dependent inhibitory), the basket cell, and chandelier cell in the brain might be candidate targets.

Abnormal inhibitory neurotransmission causes brain disorders [42, 43]. Gonzalez-Burgos et al. reported that cognitive deficits in schizophrenia may result from a GABA synapse dysfunction that disturbs neural synchrony [43]. The blockage of the antidromic invasion by GABA is not a very scientific fiction idea but physiological phenomena because it is measureable. Legitimacy of this phenomenon requires investigation in the intact brain.

Our brain can quickly shift between different conditional schemes, executing complex cognitive processes. Perhaps our brain has the switching tools to represent concepts, inner world of thought and desire, images and idea, self and consciousness, and so on. "Some interneurons might be to determine the timing of action-potential firing during rhythmic activity" (see the review by Spruston, in page 212 of ref. [44]).

5.2.4 Insight accompanied by "Aha" experience or "Eureka moment"

People solve problems with a unique process called insight, accompanied by "Aha" experience [45] and revealed a sudden burst of high-frequency gamma-band neural activity at a time of insight solution (see Figures 4–6 of ref. [45]. Both alpha-band and gamma-band power (scalp electro-encephalogram recording with power spectrum analysis) quickly shifted from a low state to a high state or vice versa at an insight moment. It took approximately 500 ms for this switching. Thus, the switching speed is ca. 500 ms.

In a study where impulse firing rate was studied [46], the switching speed of 260–610 ms was calculated (Pair #1 in Figure 5 in ref. [46] was used for this calculation), although the authors focus was not the speed of switching. In a computer model of neuron and circuit [47], the switching speed of 1.4 s was calculated (Figure 7B in ref. [47] was used for this calculation).

In **Figure 12**, the firing frequency before GABA is 1.64 Hz (antidromic spikes are invading) and that during GABA is 2.47 Hz (during the blockage). It took 500 ms for the switching. If the neuron fires much high frequency, this switching speed would be shorter.

This consideration indicates that brain wave recording in human [45], firing rate recording in rat [46], and crayfish recording (the present study **Figure 12**) do not show inconsistency.

5.2.5 Why principal neuron has a large oval soma?

Figure 13 diagrammatically illustrates principal neuron and inhibitory nerve. The dendrites integrate afferent inputs [6, 34]. Integrated currents initiate action potential at the axon. Action potential conducts both orthodromic and antidromic directions (green arrows). Electric currents necessary to initiate antidromic spikes are shown in red arrows. Inhibitory interneuron (black) makes synaptic contacts on the axon (Ax-Ax), on the soma (Soma), on proximal site of dendritic tree (p Dendrite), and on the distant site away from the soma (d Dendrite). Only one dendrite trunk is shown but it represents all dendrites, such as apical and basal dendrites of the pyramidal cell.

The diameter of the axon is 5 μm and the diameter of the oval soma is roughly 50 μm: 10 times difference in diameter between the axon and the soma. Safety factor for the action-potential generation at somal membrane significantly decreases due to less current per area. Consequently, active spike invasion tends to fail. On the other hand, "shunting switch" works efficiently at the soma area and at wider dendrite area too (see ref. [34] for dendritic integration). This consideration is the answer to the question "why principal neuron has a large oval soma?" Once again, axonal firing is relatively uneasy to stop it, because it jumps like the salutatory conduction due to a

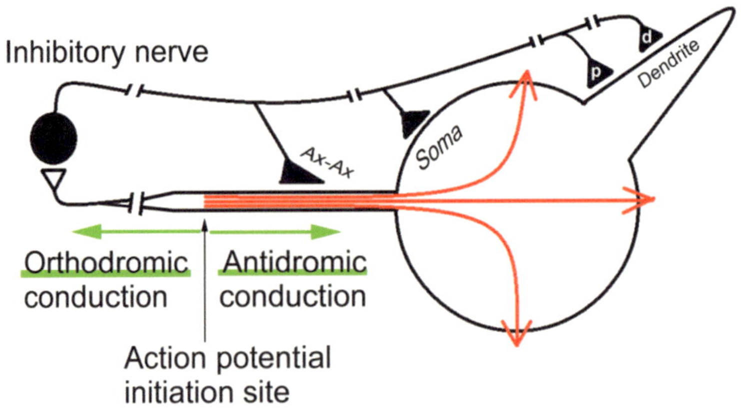

Figure 13.
Diagram to illustrate one principal neuron and one interneuron.

high safety factor of the axon. On the other hand, somal firing is relatively easy to stop, because less electric current per area flows into.

Actually, sodium channels distribute not evenly (see ref. [34]). Dendrite soma can generate Ca^{++} spikes [34]. At the dendrite, EPSP, IPSP, and presynaptic IPSP occur [34]. The dendrite is like the battle field of various currents shooting. Thus, things are not simple, but "shunting switch" scenario is untested but potential hypothesis in vertebrate brain.

Individual inhibitory interneuron has own target [3, 34]. Their target to execute shunting is genetically and behaviorally determined during maturation. Each neuron "decides" where to settle, whom to connect, what to do, finally contributes forming LTP synapses, Hebbian synapse, and/or working memory synapses, all for the sake of neuroplasticity.

Each interneuron, selectively and independently, synapses on the soma, on the "p Dendrite" or "d Dendrite" and so on. Among them, axo-axonic synapse (Ax-Ax in **Figure 13**) seems to be most powerful. It can stop all antidromic invasions. The axo-axonic synapse alignment can afford precise output control (i.e., bi-stable, on/off or 0/1 digital control).

In conclusion, in the present study, we revealed powerful switching mechanism from crustacean experiments. The switching phenomenon might be possible to contribute to various mind-mental functioning such as "Eureka," "Rubin's vase," "rabbit–duck illusion," and more such as "selective attention" [48] or "cocktail-party effect." "The science of the brain" and "the science of the mental" [49] move closer to each other. The present study was partly appeared at the IUPS international congress [50].

5.2.6 GABA in brain: Concluding remarks

Crayfish are great commercial importance as a human food delicacy [51]. The US east coast (Maine) fisheries supply 80% of the world lobster [52]. Its abundance and common have probably contributed to neurobiology progress, as Bullock [53] noted that the crustacean animals were unquestionably a giant stride forward in the understanding of substratum of nervous system function. Historically, Alexandrovicz [11, 12] first introduced lobster neuron for neurobiology research. Kuffler [21], who worked at Harvard and at Woods Hall Marine Laboratory (both closer to the US east coast), used lobster stretch receptor to bring the brain science forward, and then neurobiology has indeed advanced [54].

A small molecule, GABA, was first identified in plant extracts [55]. GABA was then found in mammalian brain tissues [56]. The report [56] showed that GABA is formed naturally from glutamic acid in the brain. However, it takes many years before GABA is recognized as an important inhibitory neurotransmitter in brain [10, 57]. At first, unknown inhibitory substance "an inhibitory factor (factor I)," was described [58], and then isolated and identified as GABA [59]. Finally, GABA formation in nerve cells and GABA release from nerve cells were experimentally demonstrated using "crustacean" nerve cells [8, 9, 22, 23].

The present study might strengthen the brain's GABA history aforementioned, that is, there is a similarity of GABAergic nervous system between invertebrate (i.e., inhibitory neurons innervating stretch receptor neurons) and vertebrate (e.g., basket interneurons innervating pyramidal neurons). In the future, "GABAergic neural switching function" that we report here, could be confirmed by electrophysiology experiments, for example, the multi-electrode recording while studying intact brain [60]. We believe that the major brain function, e.g., "how to select and match preexisting memory with events in the world [60]" could be executed by GABAergic "switching" mechanisms, which is revealed by the present study. This has worked out with heading along the way paved by Kuffler.

To conclude, a small molecule naturally occurred evolutionally very earlier (in the cells of microorganisms, plants and animals) as a non-protein amino acid [55, 61]. Meanwhile, the same molecule was utilized as a key signaling molecule in the brain of both lower and higher animals. It is beyond controversy that life on earth uses a lot of common basic molecules, like DNA, ATP, etc. to survive. In analogy with these molecules, evolutionally speaking, GABA is a fundamental substance too. Mostly, GABA works inhibitory, and thus, it helps very quietly healthy life everywhere in brain. If alterations in GABA function occur, sickness surely occurs, like epilepsy [57, 62]. This report reports the significance of such a humble substance working in our life. Healthy life is sweet indeed.

Acknowledgements

I am thankful to the Editor Dr. Noor Saher for her revisions and to Dr. G. Diarte-Plata for helpful comments. I am also thankful to T. Tsuruta, Nippon Unisoft Co., Ltd. Tokyo, for decades-long encouragements for my neurobiology study. This work is supported from the JSPS grant 21 K12688 to TY.

Author details

Toru Yazawa
Department of Biological Science, Tokyo Metropolitan University, Tokyo, Japan

*Address all correspondence to: yazawa-tohru@tmu.ac.jp

References

[1] Christopher L. Using Machine Learning to Advance Our Understanding of the Brain. BrainPost. 2021. Available from: https://www.brainpost.co/weekly-brainpost/2021/12/28/ [Accessed: Febraury 19, 2022]

[2] Cajal SR. Histologie du systeme nerveaux de l'homme et des vertebres. Paris: Maloine; 1911;**2**

[3] Eccles JC. The Understanding of the Brain. NY: McGraw-Hill; 1973, (Figures 1-4, page 7)

[4] Kuffler SW, Nicholls JG. From Neuron to Brain. The first ed. Blackwell Scientific; 1976

[5] Buzsáki G. Rhythms of the Brain. NY: Oxford University Press; 2006

[6] Stuart GJ, Spruston N. Dendritic integration: 60 years of progress. Nature Neuroscience. 2015;**18**:1713-1721. DOI: 10.1038/nn.4157

[7] Buzsáki G, da Silva FL. High frequency oscillations in the intact brain. Progress in Neurobiology. 2012;**98**:241-249. DOI: 10.1016/j.pneurobio.2012.02.004

[8] Otsuka M, Kravitz EA, Potter DD. Physiological and chemical architecture of a lobster ganglion with particular reference to gamma-aminobutyrate and glutamate. Journal of Neurophysiology. 1967;**30**:725-752

[9] Otsuka M, Iversen LL, Hall ZW, Kravitz EA. Release of gamma-aminobutyric acid from inhibitory nerves of lobster. PNAS. 1966;**56**:1110-1115

[10] Harris-Warrick R. Synaptic chemistry in single neurons: GABA is identified as an inhibitory neurotransmitter. Journal of Neurophysiology. 2005;**93**:3029-3031. DOI: 10.1152/classicessays.00026.2005

[11] Alexandrowicz JS. Muscle receptor organs in the abdomen of Homarus vulgaris and Palinurus vulgaris. Quarterly Journal of Microscopical Science. 1951;**92**:163-199. DOI: 10.1242/jcs.s3-92.18.163

[12] Alexandrowicz JS. Receptor elements in the thoracic muscles of Homarus vulgaris and Palinurus vulgaris. Quarterly Journal of Microscopical Science. 1951;**93**:315-346. DOI: 10.1242/jcs.s3-93.23.315

[13] Burgen ASV, Kuffler SW. Two inhibitory fibers forming synapses with a single nerve cell in the lobster. Nature. 1957;**180**:1490-1491. DOI: 10.1038/1801490a0

[14] Crealus W. The distribution of GABA sensitivity on crayfish muscle receptor organs. The Journal of Physiology. 1976;**263**:405-415

[15] Edwards C, Kuffler SW. The blocking effect of gamma-aminobutyric acid (GABA) and the action of related compounds on single nerve cells. Journal of Neurochemistry. 1959;**4**:19-30

[16] Edwards C, Ottoson D. The site of impulse initiation in a nerve cell of a crustacean stretch receptor. The Journal of Physiology. 1958;**143**:138-148

[17] Florey E, Florey E. Microanatomy of the abdominal stretch receptors of the crayfish (Astacusfluruviatilis L.). The Journal of Physiology. 1955;**39**:69-85

[18] Hirosawa K, Jung Hwa T-C, Nakajima Y, Tisdale AD. Thin-section and freeze-fracture studies of crayfish stretch

receptor synapses including the reciprocal inhibitory synapse. The Journal of Comparative Neurology. 1981;**200**:39-53

[19] Komuro T. Fine structural study of the abdominal muscle receptor organs of the crayfish (*Procambarusclarkii*). Morphometrical analysis of synaptic vesicles. Journal of Electron Microscopy. 1981;**30**:22-28

[20] Komuro T. Fine structural study of the abdominal muscle receptor organs of the crayfish (*Procambarusclarkii*). Sensory endings and synaptic structures. Journal of Neurocytology. 1981;**10**:27-43

[21] Kuffler SW. Excitation and inhibition in single nerve cells. Harvey Lectures. 1960;**54**:176-218

[22] Kuffler SW, Edwards C. Mechanism of gamma aminobutyric acid (GABA) action and its relation to synaptic inhibition. Journal of Neurophysiology. 1958;**21**:589-610

[23] Kuffler SW, Eyzaguirre C. Synaptic inhibition in an isolated nerve cell. The Journal of General Physiology. 1955;**39**:155-184

[24] Nadol JB Jr, de Lorenzo DAJ. Observations on the abdominal stretch receptor and the fine structure of associated axo-dendritic synapses and neuromuscular junctions in Homarus. The Journal of Comparative Neurology. 1968;**132**:419-444

[25] Nakajima Y, Tisdale AD, Henkart MP. Presynaptic inhibition at inhibitory nerve terminals. A new synapse in the crayfish stretch receptor. Proceedings of the National Academy of Sciences of the United States of America. 1973;**70**:2462-2466

[26] Ozawa S, Tsuda K. Membrane permeability change during inhibitory transmitter action in crayfish stretch receptor cell. Journal of Neurophysiology. 1973;**36**:805-816

[27] Peterson RP, Pepe FA. The fine structure of inhibitory synapse in the crayfish. The Journal of Biophysical and Biochemical Cytology. 1961;**11**:157-169

[28] Tisdale AD, Nakajima Y. Fine structure of synaptic vesicles in two types of nerve terminals in crayfish stretch receptor organs. Influence of fixation methods. The Journal of Comparative Neurology. 1976;**165**:369-386

[29] Uchizono K. Inhibitory synapses on the stretch receptor neuron of the crayfish. Nature. 1967;**214**:833-844

[30] Uchizono K. Inhibitory and excitatory synapses in vertebrate and invertebrate animals. In: Structure and Function of Inhibitory Mechanisms. Oxford: Pergamon Press; 1968. pp. 33-59

[31] Wiersma CAG, Furshpan E, Florey E. Physiological and pharmacological observations on muscle receptor organs of the crayfish, Cambarus clarkia. The Journal of Experimental Biology. 1953;**30**:136-150

[32] Yazawa T, Kuwasawa K. The cardio-regulator nerves of the hermit crabs: Anatomical and electrophysiological identification of their distribution inside the heart. Journal of Comparative Physiology. A. 1984;**154**:871-881

[33] van Harreveld A. A physiological solution for fresh water crustaceans. Society for Experimental Biology and Medicine (New York). 1936;**34**:428-432

[34] Spruston N, Stuart G, Häusser M. Principle of dendritic integration. In: Spruston N, Stuart G, Häusser M, editors. Dendrites. 3rd ed. Oxford Univ. Press; 2016. p. 736

[35] Takeuchi A, Takeuchi N. Localized action of gamma-aminobutyric acid on the crayfish muscle. The Journal of Physiology. 1965;**177**:225-238

[36] Gallo NB, Paul A, Van Aelst L. Shedding light on chandelier cell development, connectivity, and contribution to neural disorders. Tren In Neuro Science. 2020;**43**:565-580. DOI: 10.1016/j.tins.2020.05.003

[37] Schneider-Mizell CM, Bodor AL, et al. Structure and function of axo-axonic inhibition. eLife. 2021:1-39. DOI: 10.7554/eLife.73783

[38] Chiang F-K, Wallis JD, Rich EL. Cognitive strategies shift information from single neurons to populations in prefrontal cortex. Neuron. 2022;**110**:1-13. DOI: 10.1016/j.neuron.2021.11.021

[39] Que L, Lukacsovich D, Luo W, Földy C. Transcriptional and morphological profiling of parvalbumin interneuron subpopulations in the mouse hippocampus. Nature Communications. 2021;**12**:p1-p15. DOI: 10.1038/s41467-020-20328-4

[40] Hu H, Gan J, Jonas P. Fast-spiking, parvalbumin GABAergic interneurons: From cellular design to microcircuit function. Science. 2014;**345**(529):1255263-1-13. DOI: 10.1126/science.1255263

[41] Gómez CD, Rasmussen CM, Rekling JC. GABAergic inhibition of presynaptic Ca2+ transients in respiratory preBötzinger neurons in organotypic slice cultures. eNeuro. 2021;**8**:1-15. DOI: 10.1523/ENEURO.0154-21.2021

[42] Abruzzo PM, Panis C, Marini M. The alteration of chloride homeostasis/GABergic signaling in brain disorders: Could oxidative stress play a role? Antioxidants. 2021;**10**(1316):1-17. DOI: 10.3390/antiox10081316

[43] Gonzalez-Burgos G, Fish KN, Lewis DA. GABA neuron alterations, cortical circuit dysfunction and cognitive deficits in schizophrenia. Hindawi Neural Plasticity. 2011;**723184**:1-24. DOI: 10.1155/2011/723184

[44] Spruston N. Pyramidal neurons: Dendritic structure and synaptic integration. Nature Reviews. Neuroscience. 2008;**9**:206-221. DOI: 10.1038/nrn2286

[45] Jung-Beeman M, Bowden EM, Haberman J, Frymiare JL, Arambel-Liu S, Greenblatt R, et al. Neural activity when people solve verbal problems with insight. PLoS Biology. 2004;**2**:500-510. DOI: 10.1371/journal.pbio.0020097

[46] Terada S, Sakurai Y, Nakahara H, Fujisawa S. Temporal and rate coding for discrete event sequences in the hippocampus. Neuron. 2017;**94**: 1248-1262.e4

[47] Gleeson P et al. Open source brain: A collaborative resource for visualizing, analyzing, simulating, and developing standardized models of neuron and circuits. Neuron. 2019;**103**:395-411. DOI: 10.1016/j.neuron.2019.05.019

[48] Gehmacher Q, Reisinger P, Hartmann T, Keintzel T, Rösch S, Schwarz K, et al. Direct cochlear recordings in humans show a theta rhythmic modulation of auditory nerve activity by selective attention. The Journal of Neuroscience. 2022;**42**:1343-1351. DOI: 10.1523/JNEUROSCI.0665-21.2021

[49] Churchland PS. High-level affairs. Computer models of mind: Computational approaches in theoretical psychology. Book reviews. Nature. 1988;**334**:22-23

[50] Yazawa T, Ikeda K. Inhibitory transmission at the dendritic, somal, and axonal synapses on the stretch receptor neuron of the crayfish, *Procambarus clarkii*. In: The IUPS International Congress Held in Vancouver. Canada; 1986. Poster #P261.41

[51] Holdich DM. Biology of Freshwater Crayfish. Oxford: Blackwell Science; 2002

[52] Maine Lobster Marketing Collaborative. Available from: https://lobsterfrommaine.com/ [Accessed: December 21, 2022]

[53] Bullock TH. Neuron doctrine and electrophysiology. Science. 1959;**129**:997-1002

[54] Shepherd GM. Neurobiology. 3rd ed. NY: Oxford University Press; 1994

[55] Steward FC, Thompson JF, Dent CE, Gamma-Aminobutyric acid. A constituent of the potato tuber? Science. 1949;**110**:439-440

[56] Roberts E, Frankel S. γ-Aminobutyric acid in brain: Its formation from glutamic acid. The Journal of Biological Chemistry. 1950;**187**(14794689):55-63

[57] Spiering MJ. The discovery of GABA in the brain. The Journal of Biological Chemistry. 2018;**293**:19159-19160. DOI: 10.1074/jbc.CL118.006591

[58] Florey E, McLennan H. Effects of an inhibitory factor (factor I) from brain on central synaptic transmission. The Journal of Physiology. 1955;**130**:446-455

[59] Basemore AW, Elliot KA, Florey E. Isolation of factor I. Journal of Neurochemistry. 1957;**1**:334-339

[60] Buzsaki G, McKenzie S, Davachi L. Neurophysiology of remembering. Annual Review of Psychology. 2022;**73**:187-215. DOI: 10.1146/annurev-psych-021721-110002

[61] Ramos-Ruiz R, Poirot E, Flores-Mosquera M. GABA, a non-protein amino acid ubiquitous in food matrices. Cogent Food & Agriculture. 2018;**4**:1-90. DOI: 10.1080/23311932.2018.1534323

[62] Enna SJ, Gallagher JP. Biochemical and electrophysiological characteristics of mammalian GABA receptors. International Review of Neurobiology. 1983;**24**:181-212